OXFORD LOGIC GUIDES

GENERAL EDITOR : DANA SCOTT

CHOICE SEQUENCES

A CHAPTER OF INTUITIONISTIC MATHEMATICS

BY

A. S. TROELSTRA

CLARENDON PRESS · OXFORD
1977

Oxford University Press, Walton Street, Oxford OX2 6DP

OXFORD LONDON GLASGOW NEW YORK
TORONTO MELBOURNE WELLINGTON CAPE TOWN
IBADAN NAIROBI DAR ES SALAAM LUSAKA ADDIS ABABA
KUALA LUMPUR SINGAPORE JAKARTA HONG KONG TOKYO
DELHI BOMBAY CALCUTTA MADRAS KARACHI

ISBN 0 19 853163 X

For Jane Bridge

Printed in Great Britain
by Morrison & Gibb Limited, Edinburgh

PREFACE

These notes on choice sequences originated in a series of lectures given at the Mathematical Institute at Oxford during Michaelmas Term 1973. The aim of these lectures was to introduce choice sequences informally, stressing the concepts, not the purely formal axiomatic treatment. Applications and metamathematical results which might be helpful in understanding the notions were also described, but as a rule not proved, since most of the relevant literature is easily accessible.

However, in preparing the notes of these lectures for publication, it seemed useful not only to describe the relevant metamathematical results, but to include sketches of proofs as well, so as to give the reader an idea of the methods, and to facilitate the study of detailed proofs in the existing literature.

In some cases, the treatment in the literature was defective, in other cases unnecessarily formal, thereby obscuring the underlying ideas. Thus I have included (1) a rather detailed discussion of the elimination of lawless sequences (Chapter 3); (2) a simpler and more informal presentation of the continuity properties of the universe $\mathcal{U}$ and details on the connection between topological models and validity in universes $\mathcal{U}_\alpha$ (Chapter 4); (3) a modernized presentation of the main result of Dyson and Kreisel (1961) (Chapter 7).

As a result, these notes constitute a fairly comprehensive introduction to the topic of choice sequences. Chapters 1-7 contain 'what everybody with a serious interest in intuitionism should know about choice sequences'; the three appendices are devoted to historical remarks (including some discussion of the literature), illustrations of certain aspects, and 'open ends' (i.e. unfinished or inconclusive developments) of the subject which might stimulate further research.

As to the prerequisites I shall assume some general logical background (such as first order predicate logic with the completeness theorem, some basic notions from recursion theory) which may be gleaned from any not-too-elementary standard text (e.g. Kleene's Mathematical Logic), and in addition, some familiarity with the basic principles of intuitionism, such as can be obtained from any of the following sources: Heyting (1956), Chapters I-III, VII; Troelstra (1969), sections 1-8; van Dalen (1973), sections 0-2, or Dummett (1976), Chapters I-II.

Acknowledgements.

Various people have helped me by critical comments, lists of misprints etc.; I wish to mention here especially G. Kreisel, R. C. de Vrijer, and my wife. I am indebted to H. de Swart for his permission to make use of his unpublished work on the completeness of intuitionistic predicate logic for Chapter 7. My assistants G. F. van der Hoeven, G. Renardel, and J. J. Willemen assisted me in the tedious task of proof reading.

It was Jane Bridge who first suggested that the notes of my lectures on choice sequences might be suitable for publication in the Oxford Logic Guides. Her friendship greatly contributed to the enjoyment of my family and myself during our stay in Oxford, and therefore I am happy to be able to dedicate these notes to her.

Muiderberg,
August 1975. A.S.T.

CONTENTS

1
INTRODUCTORY REMARKS; PRELIMINARIES; LAWLIKE OBJECTS

1.1.

In these notes, we shall adopt the intuitionistic viewpoint, not as a philosophy of mathematics that excludes others, but as the appropriate framework for describing a *part* of mathematical experience.[†] That is to say, we shall be dealing with concepts (choice sequences) which are most naturally treated intuitionistically.

Our interpretation of 'intuitionistic' implies that we shall adopt the subjectivistic view[††] of mathematical truth: what is true is, for the idealized mathematician, what he can establish for himself by (mental) reflection about his own constructions (which include in particular those reflections (proofs)). Mathematical language is secondary; in particular, mathematical objects are not necessarily presented to us in a linguistic framework. In this respect our view agrees with that of classical set theory or geometry, where not every set of natural numbers nor every point in the plane is assumed to be definable (in a given linguistic framework). 'Secondary' means here only that the use of lan-

---

[†]The view that various parts of mathematical experience (in the widest sense) correspond to different views in the philosophy of mathematics is clearly expressed and discussed in Kreisel (1965), pages 95-98, 184-192.

[††]We use 'subjectivistic', in contrast to 'objectivistic', as referring to a concept of mathematical truth, as in platonism, which is independent of human knowledge of the truth; not in contrast to 'intersubjective' (in the sense of being valid for all mathematicians), since the 'idealized mathematician' is an idealization of any individual mathematician. But cf. also the remark on the theory of the 'creative subject' below.

guage is *mathematically* irrelevant to our objects of study, not that it is generally unimportant: in practice, language is an indispensable tool, not only because we ourselves are not idealized mathematicians with unlimited memory, but also because many special classes of operations are introduced by definability conditions (schemata).

It should be stressed at this point that talking about the 'idealized mathematician' in no way commits us to adopt the speculative features of Brouwer's theory of the 'creative subject' (the idealized mathematician); that is, first the possibility of explicit reference to the course of mathematical activity of the idealized mathematician in constructions by the idealized mathematician, and secondly the division of *all* mathematical activity into ω stages. The idealization in the concept of 'idealized mathematician' is of the same sort as that required by a platonist philosophy of mathematical knowledge, where we also assume e.g. the possibility of undistorted and immediate insight into the cumulative hierarchy of sets.

1.2. *Lawlike objects.*

A mathematical object will be called "lawlike" if it is a completed construction, something we can describe (to ourselves) completely. In some publications the word 'constructive' is used instead of lawlike (e.g. in Dragalin 1973). The various concepts of choice sequence will constitute an extension of the domain of (lawlike) mathematical objects; as will be clear from our discussions in Chapter 2 and afterwards, they cannot be considered to be completed constructions.

The two principal kinds of lawlike objects we have to consider are natural numbers and lawlike functions of natural numbers (with natural numbers as values). There is no need to say much about the natural numbers - but let us consider the idea of a lawlike function of natural numbers somewhat more closely. A lawlike function (of the type $N \to N$, N denoting the natural numbers) should be a law ('recipe') determining in an effective way a value for each argument; the law is a completely described mathematical object. Such a law is given to us in a way that makes it clear that the law is applicable to each natural number, i.e. a proof that the value is always defined is implicit. So a lawlike function can certainly not be specified by simply presenting its graph - that is a far too platonistic idea. But for example presenting a gödelnumber is not enough either - we need to know that the gödelnumber is the number of a total function, so a proof of this fact must then be supposed to be appended. Primitive recursive functions, presented in the standard way, are obviously total, and hence lawlike.

Apart from natural numbers and lawlike functions of natural numbers, there are two other categories of lawlike ob jects which we shall enoounter: species (sets of relations) and lawlike operations (functionals) defined on various universes of sequences (e.g. defined on all lawless sequences) and taking natural numbers or other sequences as values; this type of lawlike object will be discussed at length later on.

In these notes we do not commit ourselves as to which instances of the comprehension schema we consider acceptable intuitionistically - the discussion of the theory of choice sequences is to a large extent independent of the comprehension axioms adopted.

1.3. Church's thesis for lawlike functions.

Should we accept the intuitionistic form of Church's thesis, i.e. the statement

'Every lawlike function is recursive'?

Or expressed formally[†]:

CT $\quad \forall a \exists x \forall y \exists z (Txyz \;\&\; Uz = ay)$

(where a is a variable ranging over lawlike sequences, x, y, z variables ranging over N, T is Kleene's T-predicate, and U the result-extracting function as in Kleene 1952).

[†]The quantifiers are of course interpreted intuitionistically!

There are two reasons for abstaining from the identification 'lawlike = recursive':

(i) An axiomatic reason: the developments in the sequel do not depend on this identification - therefore explicitly assuming recursiveness means carrying unnecessary information around. In the formal developments, there are many possible interpretations for the range of the variables for lawlike sequences (e.g. the classical universe of sequences).

(ii) A second reason is 'philosophical': the (known) informal justifications of 'Church's thesis' all go back to Turing's conceptual analysis (or proceed along similar lines).

Turing's analysis strikes me as providing very convincing arguments for identifying 'mechanically computable' with 'recursive', but as to the identification of 'humanly computable' with 'recursive', extra assumptions are necessary which are certainly not obviously implicit in the intuitionistic (languageless) approach as adopted here. See also Gödel's remarks on pages 72-73 of Davis (1965).

1.4. *Intensional and extensional aspects.*

In Troelstra (1975) I have discussed at some length the distinction 'extensional-intensional'; there is no need to repeat the discussion here in full. For our purposes, the following will suffice: 'extensionally' in connection with functions means referring to the *graph* of the function alone. In other words, if α,β are functions, 'α is extensionally equal to β' (notation $\alpha = \beta$) is defined by

$$\alpha = \beta \equiv_{\text{def}} \forall x(\alpha x = \beta x)$$

and a predicate A (of functions) is said to be extensional if

$$\forall\alpha\forall\beta(\alpha = \beta \ \& \ A\alpha \rightarrow A\beta).$$

But of course, functions are *given* to us in some way - which means that we actually have more information about a function than just its graph. For example the gödelnumber of a recursive function b cannot be determined from the graph of b alone - although when b is given to us as recursive, the available information must permit us to find a gödelnumber for b. Another example, from classical mathematics, is provided by the treatment of functions in category theory, which are supposed to be specified with domain and codomain, not simply by their graph.

We shall loosely refer to this extra information as 'the intensional aspects', and use phrases like 'speaking intensionally', meaning: 'with reference to the intensional aspects'.

Intensional equality in the strict sense is really identity: two objects are intensionally equal if, and only if, they are given to us as the same object.

1.5. Axioms of choice, selection principles.

The following 'axiom of choice' AC_{00} or AC-NN

AC-NN $\forall x \exists y A(x,y) \rightarrow \exists a \forall x A(x,ax)$

(x,y numerical variables, a,b ranging over lawlike functions = lawlike sequences) is almost more logical than mathematical in character, i.e. it follows from the intended meaning of the logical operations: a proof of $\forall x \exists y A(x,y)$ should contain a method ('rule', 'law', 'recipe') for constructing a y to each x - but such a method is nothing else but a lawlike function.

Somewhat more generally, we have the schema AC_{01} or AC-NF:

AC-NF $\forall x \exists a A(x,a) \rightarrow \exists b \forall x A(x,(b)_x)$

where $(b)_x = \lambda y.bj(x,y)$, j is a pairing function onto the natural numbers with j_1,j_2 its inverses, and where A is supposed to be extensional in the function parameter:

$$A(x,a) \ \& \ a = a' \rightarrow A(x,a').$$

The justification is similar: a proof of $\forall x \exists a A(x,a)$ contains a method for finding an a for each x; now define b by: to compute bz, take $j_1 z$, find the a from the method given by the proof of $A(j_1 z,a)$, and let $bz = a(j_2 z)$.

We should *not* expect (even for extensional A)

(1) $\forall a \exists x A(a,x) \rightarrow \exists \Gamma \forall a A(a,\Gamma a)$

to hold (Γ an operator from functions to natural numbers) when we require Γ to be a functional in the classical extensional sense, i.e. satisfying

$$a = b \to \Gamma a = \Gamma b.$$

On the assumption of CT it is quite easy to see why (1) is implausible if Γ is to be extensional: take $\forall y \exists z(Txyz \;\&\; Uz = ay)$ for $A(a,x)$, then (1) requires an *extensional* Γ assigning a gödelnumber to each recursive a. Then Γ cannot possible be represented by a partial recursive operation defined on all gödelnumbers of total recursive functions. There is an obvious solution for Γ, namely the 'identity' function (i.e. assigning x to the recursive function with gödelnumber x), but this solution is obviously not extensional.

The only fact which is clearly implicit in the intended interpretation of the quantifier-combination $\forall a \exists b$ is that there is some operator Ψ such that $\forall a A(a, \Psi a)$; but Ψ may depend for its evaluation not only on the graph of a but also on intensional aspects of a.

For example, let $\mathfrak{A}$ range over a definable set $X = \{x: Bx\}$ of natural numbers, and y over the natural numbers N. The 'natural' equality relation on X is the equal ty induced by equality on N (i.e. equality between natural numbers); but elements of X are given to us as such by a pair (n,p), $n \in N$, p a proof of Bn. Now $\forall \mathfrak{A} \exists y A(\mathfrak{A}, y)$ becomes $\forall x(Bx \to \exists y A(x,y))$; and on the interpretation of the logical constants we see that this implies that y can be found depending on x *and* on a proof of Bx; the method for finding y is not 'extensional in x'.

Quite generally, we shall refer to principles of the form

$$\forall \mathfrak{A} \exists \mathfrak{B} A(\mathfrak{A}, \mathfrak{B}) \to \exists \Psi \forall \mathfrak{A} A(\mathfrak{A}, \Psi \mathfrak{A})$$

as 'selection principles'; it is understood that Ψ need not be extensional.

1.6. Convention.

We shall often talk about 'lawlike' sequences meaning sequences extensionally equal to lawlike ones (but not necessarily given as a lawlike sequence). This will not be a prolem as most of the time we shall discuss extensional properties of sequences only.

1.7. *Choice sequences as a generalization of the concept of sequence.*

Lawlike sequences are completely determined by a law given in advance. What is essential for mathematical purposes however, is that the concept of sequence is such that to each argument a value can be determined. The various concepts of choice sequence generalize the concept of lawlike function such that the idea of being determined by a law given in advance is abandoned, but the essential property of a value existing for each argument is retained.

The simplest concept of choice sequence is that of a lawless sequence, which will be discussed extensively in Chapter 2. For lawless sequences we can present an axiomatization completely characterizing their properties relative to a theory of lawlike objects; the axiomatization is obtained by a very convincing conceptual analysis. For the philosophy of mathematics, the possibility of such an analysis seems to us to be of considerable interest; and this interest is not changed by the fact that the results show that quantification over lawless sequences may be viewed as a 'manner of speech'.

1.8. *Description of* EL *and* EL_1.

EL (EL = elementary analysis) contains variables for natural numbers: (x,y,z,u,v,w,n,m) and for (lawlike) functions (a,b,c,d); constants: 0 (zero), S (successor), = (equality between natural numbers), λ (abstraction operator for the introduction of functions by explicit definition), R (recursor, an operation for definition by recursion), j,j_1,j_2 (a pairing function onto the natural numbers with inverses), and Φ (for application of functions to numbers); the logical constants are &, $\vee$, $\rightarrow$, $\forall$, $\exists$ (for functions and numbers).

Below we shall use A, B, C, D as syntactical variables for formulae, t, s as syntactical variables for numerical terms, and ϕ, ψ, ξ for functors (functional terms). $\neg A$ is an abbreviation of $A \to S0 = 0$; logical equivalence $\leftrightarrow$ is also treated as defined. ϕt abbreviates $\Phi\phi t$. Other conventions and abbreviations which will not be explained are standard (e.g. 1, 2, 3, ... for $S0$, $SS0$, $SSS0$, ...).

To indicate parameters in terms and functors we use square brackets; e.g. $t[x]$ indicates a term with (numerical) parameter x. Once $t[x]$ has been introduced, $t[t']$ indicates the term obtained by substitution of t' for x in t; a more accurate, but also more cumbersome notation is $t_x[t']$.

$\underset{\sim}{\mathrm{EL}}$ is based on two-sorted intuitionistic predicate logic, contains the usual axioms for successor and equality, pairing axioms:

$$j_1 j(x,y) = x,\ j_2 j(x,y) = y,\ j(j_1 z, j_2 z) = z,$$

induction with respect to all formulae in the language, the conversion rule

$$(\lambda x.t[x])t' = t[t']$$

(possibly renaming bound variables in t so as to avoid clashes of variables), the axioms for primitive recursion

$$\begin{cases} Rxa0 = x \\ Rxa(Sy) = aj(Rxay, y) \end{cases}$$

and a very weak axiom of choice

QF-AC $\quad \forall x \exists y A(x,y) \to \exists a \forall x A(x,ax) \quad$ (A quantifier-free).

In the presence of primitive recursive functions, QF-AC expresses closure under 'recursive in' for the domain of functions, therefore the minimal model for QF-AC consists of the natural numbers and all recursive functions over the natural numbers.

$\underset{\sim}{\mathrm{EL}}_1$ is the system obtained from $\underset{\sim}{\mathrm{EL}}$ by replacing QF-AC by

AC-NF $\quad \forall x \exists a A(x,a) \to \exists b \forall x A(x,(b)_x)$

where, as before,

$$(b)_x \equiv_{\mathrm{def}} \lambda y.bj(x,y).$$

1.9. Other notations and conventions (for consultation when needed).

For future use we introduce here some other notations and conventions.

(A) For coding u-tuples we use ν_u, with inverses $j_1^u, j_2^u, \ldots, j_u^u$:

$$\nu_u(j_1^u x, \ldots, j_u^u x) = x, \; j_i^u \nu_u(x_1, \ldots, x_u) = x_i.$$

(For the sake of definiteness, we may take e.g.

$$\nu_1(x) = x, \; \nu_2(x,y) = j(x,y), \; \nu_{u+1}(x_1, \ldots, x_{u+1}) = j(\nu_u(x_1, \ldots, x_u), \; x_{u+1}).)$$

Also $b(x_1, \ldots, x_u)$ stands for $b\nu_u(x_1, \ldots, x_u)$.

(B) We assume a (primitive recursive) coding of all finite sequences of natural numbers *onto* the natural numbers to be given; in talking about sequences, we usually shall not distinguish between the sequence and its coding. Let $\langle x_0, \ldots, x_u \rangle$ be the (code number of) the sequence $x_0, \ldots, x_u$; $*$ indicates *concatenation*, i.e.

$$\langle x_0, \ldots, x_u \rangle * \langle x_{u+1}, \ldots, x_v \rangle = \langle x_0, \ldots, x_v \rangle.$$

We shall assume

$$\langle\rangle = 0.$$

We put

$$n \preceq m \equiv_{\text{def}} \exists n'(n * n' = m)$$
$$n \prec m \equiv_{\text{def}} n \preceq m \; \& \; n \neq m.$$

The *length-function* lth satisfies

$$\begin{cases} \text{lth} \langle\rangle = \\ \text{lth} \langle x_0, \ldots, x_u \rangle = u+1. \end{cases}$$

For the *inverse* to sequence-coding we write $(n)_x$, i.e. if $n = \langle x_0, \ldots, x_u \rangle$, then

$$(n)_y = x_y \text{ for } y \leq u, \; (n)_y = 0 \text{ elsewhere.}$$

We introduce the abbreviation

$$\hat{n} \equiv \langle n \rangle$$

and

$$\begin{cases} \bar{a}0 \equiv \langle\rangle \\ \bar{a}u \equiv \langle a0, \ldots, a(u \dot{-} 1) \rangle \end{cases}$$

(where $x \dot{-} y$ is defined as usual by $x \dot{-} 0 = x$, $x \dot{-} Sy = \text{prd}(x \dot{-} y)$, and prd (predecessor) satisfying $\text{prd}\; 0 = 0$, $\text{prd}(Sx) = x$). Note that $\bar{\alpha}(u+1) = \bar{\alpha}u * \langle \alpha u \rangle$.

Finally we put

$$a \in n \equiv_{\mathrm{def}} \overline{a}(\mathrm{lth}(n))=n \equiv \exists y(\overline{a}y = n).$$

For the purpose of conceiving sequences as sequences of p-tuples we need $k_1^p,\ldots,k_p^p$, satisfying

$$\begin{cases} k_i^p\, 0 = 0 \\ k_i^p(n*\hat{x}) = k_i^p n * \langle j_i^p x\rangle \end{cases} \qquad i = 1,\ldots,p.$$

2
LAWLESS SEQUENCES

2.1. Introduction.

Chapters 2-4 in this monograph are devoted to lawless sequences. They claim our interest for the following reasons:
(a) Lawless sequences provide a simple example of a concept for which, on the basis of a *precise* informal description,we can derive informally ('justify') the axioms stated for this concept; moreover, the most natural justification is presented in subjectivistic terms. Acceptance of lawless sequences as mathematical objects implies denial of the thesis that all mathematical entities should be given to us via a linguistic representation. This point is elaborated in the present chapter.
(b) It is possible to treat quantification over lawless sequences as a 'figure of speech'; this follows from the elimination theorems (Chapter 3).
(c) Lawless sequences provide the starting point for the construction of universes of sequences which satisfy some non-trivial continuity schemata and are closed under lawlike continuous operations; these universes thereby become suitable as a basis for (intuitionistic versions of) 'traditional' mathematical theories such as real-valued analysis, while adding some 'distinctive flavour' to these theories owing to the validity of continuity schemata (Chapter 4).
(d) They illustrate the influence of mathematical assumptions on the extent of the concept of 'validity in all structures' for formulae of intuitionistic predicate logic **IPC**, and provide some motivation for the choice of the set of theorems of **IPC**. Moreover, lawless sequences enable us to equate validity in Beth models with validity in intuitionistic structures (modulo certain assumptions; see Chapter 7).

2.2.

Lawless sequences provide the simplest example of a concept of sequences where the sequences are not thought of as being completely determined in advance by a law. We think of a lawless sequence (of natural numbers),as a process (not a law!) of assigning values to the arguments 0, 1, 2 ...; at any stage, only finitely many values (i.e. an initial segment) of the sequence are known; at no stage we impose restrictions on future possibilities for assigning values to arguments (except, in this case, the general a priori restriction that all values will be natural numbers). So a lawless sequence is certainly a sequence, inasmuch as to each argument eventually a value will be assigned. But we have 'abstracted' from the idea of a law determining the assignment of values to arguments in advance. A good analogy ('model') is provided by the sequence of casts of a die; at any stage only finitely many results are known, but we cannot say anything about future values (except, in this case, that all values belong to {1,2,3,4,5,6}). The analogy becomes even better, if we permit at the beginning a finite number of deliberate placings of the die, so as to ensure that any finite sequence of values taken from the set {1,...,6} can occur as initial segment of the sequence. (We wish to ensure that the lawless sequences are dense in all number-theoretic functions, i.e. all initial segments do occur; cf. section 2.4 below.)

So far we have avoided the terminology of 'choice' and 'to choose' which has been customary in traditional intuitionistic literature. This terminology has often given rise to objections of the type: 'to talk about (free) choices is introducing an element of arbitrariness (subjectiveness) into mathematics, where it does not belong'. But what is relevant from a mathematical point of view is not any individual choice sequence as such, but the 'mathematical' fact that there exist many perfectly well-defined (lawlike) operations on sequences which can be carried out without assuming the arguments to be determined by a law. Example: $\alpha+\beta \equiv_{\mathrm{def}} \lambda x.(\alpha x+\beta x)$ constructs a new sequence out of two given ones. As we shall see below, if α,β are lawless, $\alpha+\beta$ is again a sequence, but not a lawless one.

In order to avoid possible objections or fruitless speculations, let us stress that of course again certain idealizations enter in the formation of the concept of a lawless sequence: one 'abstracts' from 'accidental' features of such a process, such as the length of the time-interval between successive choices, or at which moment the process was started. There is also no mystery in the use of the term 'abstracting' in the sense of 'forgetting deliberately'; this is done quite often in mathematics: we 'forget' additional structure when it is not relevant to a problem. We shall return to this matter later.

2.3. Convention.

In the remainder of this section, we shall reserve α, β, γ, δ for lawless sequences, a, b, c, d as before for lawlike sequences, and ξ, η, ζ for arbitrary sequences.

For the arguments (sequences of natural numbers, species of natural numbers etc.) of the predicates we shall consider, we assume a notion of extensional equality to be defined; the predicates are always assumed to be extensional in their arguments. This condition is in fact redundant for lawless sequences as arguments. Further, in the discussion of the axiom schemata below we shall assume the lawless sequences to be the only non-lawlike parameters occurring; but this restriction may be relaxed, cf. section 2.17.

2.4. Existence axiom specifying initial segments.

The first axiom we propose is

LS1 $\quad \forall n \exists \alpha (\alpha \in n)$.

(For any given initial segment we can find a lawless sequence starting with that initial segment.) This axiom makes the description of lawless sequences in informal terms, as in section 2.1, slightly more specific: we may specify an initial segment in advance, or decide to assign values to a finite set of arguments so as to conform to a prescribed initial segment - but we are not permitted to make general restrictions on all future choices of values - if we do so, we are thinking not of a lawless sequence, but of another concept. This axiom also shows that the analogy to the sequence of casts of a die is not quite accurate, unless we allow for a finite number of deliberate placings of the die at the beginning.

2.5. *Decidability of equality and identity.*

Let $\equiv$ denote (intensional) identity between lawless sequences: $\alpha \equiv \beta$ means that α, β are (given to us as) the *same* process of generating values. Identity is obviously decidable:

(1) $\alpha \equiv \beta \vee \alpha \not\equiv \beta$,

i.e. we think of α and β as either the same process or as distinct processes.

If we define $\alpha = \beta$ as $\forall x(\alpha x = \beta x)$, we can show that $\alpha = \beta \leftrightarrow \alpha \equiv \beta$. The implication from the right to the left is trivial. Suppose conversely $\alpha = \beta$. How can we *know* that $\alpha x = \beta x$ for all x? Only by knowing that $\alpha \equiv \beta$; for if α, β are distinct processes, then at any stage we know at most initial segments of α, β, and even if those segments are compatible, they may have distinct incompatible continuations. So, avoiding $\equiv$ as a primitive in the statement of our schemata, we may regard $\equiv$ between lawless sequences as being *defined* by extensional equality, and we can render the content of (1) as

LS2 $\alpha = \beta \vee \alpha \neq \beta$.

2.6 - 2.8. *The axiom schema of open data.*

2.6. The schema of open data in its simplest form can be stated as follows:

(1) $A\alpha \to \exists n[\alpha \in n \ \& \ \forall \beta \in n A\beta]$

($A\alpha$ not containing other non-lawlike parameters besides α).

In the discussion below, we shall assume our predicates of lawless sequences not to contain other non-lawlike parameters.

Justification: Suppose we are able to assert $A\alpha$ (i.e. we have a proof of A for α; we have this proof at a certain stage in the construction of α). At any stage we only know an initial segment of α; so $A\alpha$ was also asserted on the information that $\alpha \in n$. Therefore A should hold for any lawless sequence starting with n.

Now we wish to generalize (1) to the case where A contains other lawless parameters besides α. As a generalization

(2) $\quad A(\alpha,\beta,\ldots) \rightarrow \exists n[\alpha \in n \ \& \ \forall \gamma \in n A(\gamma,\beta,\ldots)]$

is incorrect. For a counterexample, take $A(\alpha,\beta) \equiv \alpha=\beta$; then the conclusion of (2) becomes obviously false:

$$\exists n[\alpha \in n \ \& \ \forall \gamma \in n(\alpha=\gamma)].$$

For any n, this is refuted by choosing $\gamma \in n*\langle\alpha(\mathrm{lth}(n))+1\rangle$.

Before we state the correct generalization of (1), however, we first introduce some notational conventions.

2.7. *Notations and conventions.*

$$\neq(\alpha,\beta_1,\ldots,\beta_n) \equiv_{\mathrm{def}} \alpha\neq\beta_1 \ \& \ \alpha\neq\beta_2 \ \& \ \ldots \ \& \ \alpha\neq\beta_n,$$

$$\#(\beta_0,\beta_1,\ldots,\beta_n) \equiv_{\mathrm{def}} \neq(\beta_0,\beta_1,\ldots,\beta_n) \ \& \ \neq(\beta_1,\beta_2,\ldots,\beta_n) \ \& \ \ldots \ \& \ \neq(\beta_{n-1},\beta_n),$$

or equivalently

$$\#(\beta_0,\ldots,\beta_n) \equiv_{\mathrm{def}} \bigwedge_{0\le i<j\le n} \beta_i\neq\beta_j.$$

General convention for the remainder of the book: in expressions no lawless parameters are assumed to occur besides those explicitly shown.

$$\underline{\forall}\alpha A(\alpha,\beta_1,\ldots,\beta_n) \equiv_{\mathrm{def}} \forall\alpha(\neq(\alpha,\beta_1,\ldots,\beta_n) \rightarrow A)$$

$$\underline{\exists}\alpha A(\alpha,\beta_1,\ldots,\beta_n) \equiv_{\mathrm{def}} \exists\alpha(\neq(\alpha,\beta_1,\ldots,\beta_n) \ \& \ A).$$

Note that

$$\underline{\forall}\alpha_1 \ldots \underline{\forall}\alpha_p A(\alpha_1,\ldots,\alpha_p) \leftrightarrow \forall\alpha_1\ldots\forall\alpha_p(\#(\alpha_1 \ldots \alpha_p) \rightarrow A(\alpha_1 \ldots \alpha_p))$$

and similarly for $\underline{\exists}\alpha_1\ldots\underline{\exists}\alpha_p A(\alpha_1,\ldots,\alpha_p)$.

Abbreviations such as $\underline{\forall}\alpha\in nA(\alpha)$, $\underline{\exists}\alpha\in nA(\alpha)$ are self-evident.

2.8.

Now we are able to state in a concise form the correct generalization of (1) of section 2.6:

LS3 $A(\alpha,\beta_1,\ldots,\beta_n)$ & $\neq(\alpha,\beta_1,\ldots,\beta_n) \rightarrow$
$\rightarrow \exists n[\alpha\in n$ & $\underline{\forall}\gamma\in nA(\gamma,\beta_1,\ldots,\beta_n)]$.

Comment. LS2 and LS3 force us to consider possibly distinct lawless sequences simultaneously. It is correct to say, as long as we are discussing a single lawless sequence only, that at any stage all information is contained in a single initial segment. But when talking about different lawless sequences, this might lead us to suppose that sequences of which, at a given stage, the same initial segment is known, are really indistinguishable. But of course, that is not what is intended: each process (= lawless sequence) has, apart from a known initial segment at any stage, also its 'individuality' (i.e. its identity as a process); and different individuals from the universe of lawless sequences, even if the same initial segment is known for them at a certain stage, may be continued differently.

In this connection, the following example may have some pedagogic value. Suppose we have started two lawless sequences α and β, alternately selecting values: $\alpha 0$, $\beta 0$, $\alpha 1$, $\beta 1$, $\alpha 2$, $\beta 2$, ... Now we may also regard this as a *single* process γ, with $\gamma(2n) = \alpha n$, $\gamma(2n+1) = \beta n$. However, we cannot regard α, β, γ as all being lawless within the same context: either we have to decide α and β to be lawless, and then γ is a sequence constructed from α,β which is not itself lawless (since the information available about γ at any stage is of a kind not allowed for a lawless sequence: γ is completely determined relative to α,β); or we choose to consider γ as lawless, in which case α,β are sequences (not lawless ones) constructed from γ.

Formally, we can show that γ,α,β cannot be simultaneously lawless. We may assume $\alpha\neq\beta$ (since this was intended in our construction of γ). $\gamma=\alpha$ would yield $\gamma(2x) = \alpha(2x) = \alpha x$, hence also $\alpha(2^n) = \alpha(1)$ for all n. But then (1) of section 2.6 yields

$$\forall n(\alpha(2^n) = \alpha(1)) \rightarrow \exists m[\alpha\in m \ \& \ \forall\beta\in m\forall n(\beta(2^n) = \beta(1))]$$

and the conclusion is obviously false. Similarly $\gamma\neq\beta$. Hence $\neq(\gamma,\alpha,\beta)$ is satisfied. Now assume

$$\forall x(\gamma(2x) = \alpha x \ \& \ \gamma(2x+1) = \beta x).$$

Application of LS3 yields

$$\exists n[\gamma\in n \ \& \ \underline{\forall}\delta\in n\forall x(\delta(2x)=\alpha x \ \& \ \delta(2x+1)=\beta x)]$$

which is again obviously false.

LS3, especially in its simplest form (1) of section 2.6 is appropriately named as the schema of open data, since it can be read as: whenever $A\alpha$ holds for a lawless sequence α, it holds for an open set (in N^N restricted to lawless sequences) to which α belongs (in fact, a clopen neighbourhood of α).

2.9 - 2.15. Continuity schemata.

2.9. Continuity implicit in LS3; *stronger forms of continuity.*

Now we turn to continuity schemata. Some weak form of continuity is already implicit in LS3, especially in its simplest form (1) of section 2.6 above, since it follows from (1) of section 2.6 that

(1) $\forall\alpha\exists xA(\alpha,x) \rightarrow \forall\alpha\exists x\exists y\forall\beta\in\overline{\alpha}y\ A(\beta,x).$

To see this, assume $A(\alpha,x)$; then we find by (1) of section 2.6 a y such that $\alpha\in\overline{\alpha}y$ (trivially) and $\forall\beta\in\overline{\alpha}y\ A(\beta,x)$.

However, it is possible to express continuity in a stronger way. Let us use '$\mathrm{Cont}_{\mathrm{LS}}$' to denote the class of *lawlike* operations on lawless sequences, assigning natural numbers to lawlike sequences such that for $\Gamma \in \mathrm{Cont}_{\mathrm{LS}}$

(2) $\forall\alpha\exists x\forall\beta\in\overline{\alpha}x(\Gamma\alpha = \Gamma\beta).$

Then we can express continuity as

(3) $\forall\alpha\exists xA(\alpha,x) \rightarrow \exists\Gamma \in \mathrm{Cont}_{\mathrm{LS}}\ \forall\alpha A(\alpha,\Gamma\alpha).$

There are a number of comments to be made.

Remark A. The intended intuitionistic interpretation of the quantifier combination $\forall\alpha\exists x$ leads to an 'axiom of choice' or 'selection principle'

(4) $\forall\alpha\exists xA(\alpha,x) \rightarrow \exists\Psi\forall\alpha A(\alpha,\Psi\alpha).$

The *extra* strength of (3) over (4) is in the fact that Ψ is also assumed to be *continuous*, and this seems to be justified in view of the fact that Ψ must assign a *complete* object (i.c. a natural number); thus, for each α, $\Psi\alpha$ must be (com-

pletely) known at a certain stage, i.e. $\Psi\alpha$ is computed from an initial segment of α only.
More formally, we may apply LS3:

$$\Psi\alpha = x \rightarrow \exists n[\alpha \in n \;\&\; \forall \beta \in n(\Psi\beta = x)]$$

which implies $\Psi \in \mathrm{Cont}_{\mathrm{LS}}$. So, actually, (3) may be viewed as a combination of LS3 (in its simplest form) with the selection principle (4).

Remark B. The same type of reasoning will hold with respect to any assignment of lawlike objects ('completed' objects) to lawless sequences, the operators giving the assignment in virtue of a selection principle of type (4) being continuous as in (2).
So we have, in particular

(5) $\quad \forall\alpha\exists a A(\alpha,a) \rightarrow \exists\Gamma\in\mathrm{Cont}^1\forall\alpha A(\alpha,\Gamma\alpha)$

where 'Cont ' denotes the class of continuous functionals on LS (the universe of lawless sequences) with lawlike sequences as values. In fact, if we accept Church's thesis in the form 'lawlike = recursive', (5) reduces to a special instance of (3).

Remark C. Continuous functionals of type $F \rightarrow N$, where F is a universe of functions, can be represented by neighbourhood functions of type $N \rightarrow N$, forming a class K_F which is defined by

$$K_F(a) \equiv \forall\xi\exists x(a(\bar{\xi}x) \neq 0) \;\&\; \forall nm(an \neq 0 \rightarrow a(n*m) = an)$$

where ξ is a variable ranging over F. Intuitively, $a(\bar{\xi}x) = 0$ signifies: $\bar{\xi}x$ is too short as an initial segment for computing $\Phi_a(\xi)$ (Φ_a denoting the functional represented by a), and $a(\bar{\xi}x) = y + 1$ means $\Phi_a(\xi) = y$. The second conjunct in the defining clause of K_F is a matter of convenience, it might for example have been replaced by

$$\forall nm(an \neq 0 \;\&\; (m \prec n \vee m \succ n) \rightarrow am = 0)$$

or deleted altogether; in the latter case, we must specify that $a(\bar{\xi}x) = y+1 \;\&\; \forall x'<x(a(\bar{\xi}x) = 0)$ means $\Phi_a(\xi) = y$.

Thus we may rephrase (3) as:

(6) $\quad \forall\alpha\exists x A(\alpha,x) \rightarrow \exists a\in K_F\forall n(an \neq 0 \rightarrow \forall\alpha\in n A(\alpha,an\dot{-}1))$

(checking the equivalence is left to the reader).

If we introduce the notation $a(\alpha)$ with specification

$$a(\alpha) = x \leftrightarrow \exists y(a(\overline{\alpha}y) = x+1)$$

then we can keep closer to (3):

(7) $\forall\alpha\exists x A(\alpha,x) \rightarrow \exists a \in K_F \forall\alpha A(\alpha,a(\alpha))$.

For continuous functionals of type $F \rightarrow C$, where C is a species of lawlike objects, we may express continuity by

$$\forall\xi\exists\mathfrak{B}A(\xi,\mathfrak{B}) \rightarrow \exists a \in K_F \exists\mathfrak{M}\forall n(an \neq 0 \rightarrow \forall\xi \in n A(\xi,\mathfrak{M}(an \dot{-} 1)))$$

where $\mathfrak{M}$ is of type $N \rightarrow C$, $\mathfrak{B}$ a variable ranging over C, and ξ a variable ranging over F. Because of AC_{01} for lawlike functions, (5) may be expressed as

(8) $\forall\alpha\exists a A(\alpha,a) \rightarrow \exists a \in K_{LS}\ \exists b\forall\alpha A(\alpha,(b)_{a(\alpha)})$

or

(8') $\forall\alpha\exists a A(\alpha,a) \rightarrow \exists a \in K_{LS}\ \exists b\forall n(an \neq 0 \rightarrow \forall\alpha \in n A(\alpha,(b)_{an \dot{-} 1}))$.

Here as before $(b)_x = \lambda y.b(x,y)$.

2.10. *Continuity postulates in more general situations; functional character.*

For lawless sequences, the combination of the selection principle (4) of section 2.9 with LS3 is quite straightforward: Ψ can only act on initial segments, since at any stage no other information is available about lawless sequences.

The situation becomes more complicated however, if we are considering notions of sequence for which at any stage other information may also be available, because then Ψ may be computed not only from the numerical values of a sequence but also from the additional information; so it may happen that $\alpha = \beta$, but $\Psi\alpha \neq \Psi\beta$. In other words, Ψ does not necessarily have functional character on such sequences with respect to *extensional* equality.

The occurrence of operations 'without functional character' is not at all 'deep', and quite common: functions on real number generators (Cauchy-sequences of rationals) are not necessarily functions on the reals (which are defined as equivalence classes of real number generators), or even more trivial: rationals may be introduced as (equivalence classes of) pairs of integers, but a function defined on pairs of integers is not necessarily a function on the rationals. An example from recursion theory: the operation which assigns x to a total recursive function $\{x\}$ is not a functional with respect to extensional equality. Cf. our remarks in section 1.5.

From this it will become clear that for more general concepts of choice sequence there may be quite a difference as to intuitive justification between continuity principles for the quantifier combinations $\forall\alpha\exists x$ and $\forall\alpha\exists!x$ when applied to $A(\alpha,x)$ extensional in α. In the latter, $\exists!x$ refers to uniqueness with respect to extensional equality, i.e.

$$A(\alpha,x) \ \& \ A(\beta,y) \ \& \ \alpha=\beta \rightarrow x=y.$$

(This is in fact a consequence of uniqueness with respect to identity and extensionality of $A(\alpha,x) \ \& \ A(\alpha,y) \rightarrow x=y$, $A(\alpha,x) \ \& \ \alpha=\beta \rightarrow A(\beta,x)$, $A(\beta,y) \ \& \ \alpha=\beta \rightarrow A(\alpha,y)$, hence $A(\alpha,x) \ \& \ A(\beta,y) \ \& \ \alpha=\beta \rightarrow x=y$.) We shall meet some illustrative examples of the difference between $\forall\alpha\exists x$ and $\forall\alpha\exists!x$-continuity in Appendix B.

2.11. *More information about K_{LS} and* $\mathrm{Cont}_{\mathrm{LS}}$*: the extension principle.*

Let $\mathrm{Cont}_{\mathrm{LS}}$ denote the continuous functionals defined on LS with numerical values. Can we obtain still more information about $\mathrm{Cont}_{\mathrm{LS}}$ and K_{LS}? In this section we shall try to justify the *extension principle*:

(1) $\begin{cases}\text{Each } \Gamma \in \mathrm{Cont}_{\mathrm{LS}} \text{ can be extended to a continuous} \\ \text{operation defined on all sequences.}\end{cases}$

With the help of neighbourhood functions we may reformulate (1) as

(2) $$a \in K_{\mathrm{LS}} \rightarrow \forall\xi\exists x(a(\overline{\xi}x) \neq 0)$$

(ξ ranging over all possible sequences).

We may give a justification for this extension principle as follows: consider any sequence ξ. At any stage n, we may know, besides $\xi 0$, $\xi 1$, ..., ξn, further information R_n about ξ (a very special case would be the case where already R_0 would contain the specification of a law for ξ).

Now we apply a process 'Abstr' to ξ: at each stage, we

consider $\xi 0, \ldots, \xi n$, but 'forget' about R_n, and we attempt to compute $\Gamma\xi$ for a given $\Gamma \in \mathrm{Cont}_{LS}$, i.e. we try to find out whether $a(<\xi 0,\ldots,\xi n>) \neq 0$ if a is the neighbourhood function representing Γ. The point is that we may think of $\xi 0, \ldots, \xi n$ at any stage as representing the initial segment of some lawless sequence, and our contention that (1) (or equivalently (2)) holds is equivalent to saying that Γ 'does not know' the difference between a genuine lawless sequence and a sequence obtained from a given sequence by 'forgetting' the additional information. We may think of 'Abstr' as a kind of conceptual transformation: we attempt to view ξ as a lawless sequence Abstr(ξ) by 'forgetting' other information - in other words, when trying to apply an operation to Abstr(ξ), we permit reference only to the values of ξ. Note that 'Abstr' cannot be viewed as a mathematical operation in the ordinary sense - we cannot prove, for example, that b = Abstr(b) since, in thinking of Abstr(b) in a mathematical context, Abstr(b) in its quality of imitation-lawless sequence by its very nature cannot be assumed to be related to a lawlike sequence b by b = Abstr(b), because this is a type of information not permissible if we think of Abstr(b) as lawless. (Yet, in another 'meta'sense b = Abstr(b)).

Further reflection will reveal that there is a residue of a new assumption in the acceptance of (1) (i.e. it is not completely explained in terms of insights already accepted) but we contend that the preceding considerations make (1) extremely natural and convincing.

2.12. 'Brouwer's dogma' on the inductively defined functionals.

L. E. J. Brouwer has (in Brouwer (1927) for example) advocated a hypothesis which is much stronger than the extension principle (not for Cont_{LS} to be precise, but for the corresponding class $\mathrm{Cont}_{\mathcal{U}}$ for his universe $\mathcal{U}$ of choice sequences. However, the reasoning for $\mathrm{Cont}_{\mathcal{U}}$ should apply equally well, if not better, to Cont_{LS}).

We wrote 'advocated a hypothesis...': in fact, Brouwer

claimed to have a proof. We shall return to this claim in a moment - but first we must state the hypothesis and see what it implies. Let Φ, Φ',... stand for functionals from sequences to numbers, i.e. of type $N^N \to N$. We define Φ_n by

$$\Phi_n(\xi) = \Phi(n*\xi)$$

where $n*\xi$ is defined in the obvious way by

$$\begin{cases}(n*\xi)(u) = (n)_u \text{ if } u < \mathrm{lth}(n) \\ (n*\xi)(u) = \xi(u \dot{-} \mathrm{lth}(n)) \text{ if } u \geq \mathrm{lth}(n)\end{cases}$$

(i.e. $n*\xi$ is the concatenation of the finite sequence $\langle (n)_0,\ldots,(n)_{\mathrm{lth}(n)\dot{-}1}\rangle$ with $\xi 0, \xi 1, \xi 2, \ldots$).

Now we introduce the class Ind of functionals inductively defined by two closure conditions

(i) $\lambda\xi.n \in \mathrm{Ind}$

(ii) If $\Phi^{(0)}, \Phi^{(1)}, \Phi^{(2)}$ ... all belong to Ind, then also Φ defined by $\Phi(\alpha) = \Phi^{(\alpha 0)}(\lambda x.\alpha(x+1))$. I.e. to compute $\Phi(\alpha)$, take $\alpha 0$, and compute $\Phi^{(\alpha 0)}$ on the remainder of the sequence α.

This amounts to:

$$\forall x(\Phi_{\langle x\rangle} \in \mathrm{Ind}) \to \Phi \in \mathrm{Ind}.$$

All elements of Ind must be shown to belong to Ind by use of (i) and (ii) (hence, in classical terms, Ind is to be the minimal class X satisfying $\forall n(\lambda\xi.n\in X)$ and $\forall x(\Phi_{\langle x\rangle}\in X)\to\Phi\in X)$, which gives rise to a principle of induction over Ind:

$$\lambda\xi.n\in X \;\&\; [\forall x(\Phi_{\langle x\rangle}\in X) \to \Phi\in X] \to \mathrm{Ind} \subset X.$$

Let us use Cont for the class of functionals defined and continuous for all sequences. Then we can show

(1) $\mathrm{Ind} \subset \mathrm{Cont} \subset \mathrm{Cont}_{LS}$

by an easy induction over Ind (left to the reader). More precisely, the inductive definition of Ind selects a subclass of any class of functionals which is closed under (i) and (ii), and the proof of Ind $\subset$ Cont applies to all such subclasses. Brouwer's hypothesis would amount to

(2) $\mathrm{Cont} \subset \mathrm{Ind}$

or more specifically

(3) $\mathrm{Cont}_{LS} \subset \mathrm{Ind}$

(which actually amounts to (2) by (1) of section 2.11; note that (1) and (3) together imply (1) of section 2.11).

Accepting (3) (or (2) for that matter) requires accepting a *new* insight. Let us first indicate how the so-called bar theorem, which is usually formulated as a schema

$$\mathrm{BI}_D \qquad [\forall n(Pn \vee \neg Pn) \ \& \ \forall\alpha\exists x P(\overline{\alpha}x) \ \& \ \& \ \forall n(Pn \rightarrow Qn) \ \& \ \forall n(\forall y Q(n*\hat{y}) \rightarrow Qn)] \rightarrow Q0$$

implies (informally) (2): just take for P and Q

$Pn \equiv$ 'n is sufficiently long to compute $\Phi\alpha$ for $\alpha\in n$'

$Qn \equiv \Phi_n \in \mathrm{Ind}$,

where Φ is any functional of Cont_{LS}. It is obvious that the premise of BI_D is satisfied for these P,Q (the decidability of P follows from the fact that the algorithm determining Φ should be such that we are not in doubt whether a segment is long enough to carry through the computation), and $Q0$ just means that $\Phi \in \mathrm{Ind}$.

Kleene (in Kleene and Vesley 1965) formulates the intuitive content of the bar theorem as follows. If we think of number-theoretic functions as branches of an infinitely branching tree (Baire space)

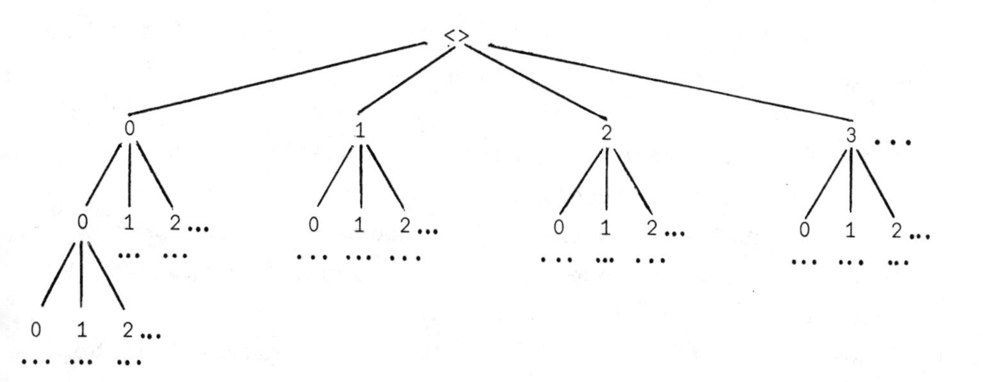

and we define 'n is explicitly securable' by

$\forall\alpha\exists x P(n*\overline{\alpha}x)$

and 'n is inductively securable' meaning n is a member of the class Q inductively defined by

$Pn \rightarrow Qn, \ \forall n(\forall y Q(n*\hat{y}) \rightarrow Qn)$,

then BI_D says that the concepts of explicit securability and inductive securability coincide. There seems to be no approach known which reduces this complex intuitive insight to essentially simpler ones.

To be plausible at all, it is certainly necessary that $\forall\alpha\exists x P(n*\overline{\alpha}x)$ can be established using information about *initial* segments only; no reference *in general*[†] should be possible to intensional aspects of functions such as gödelnumbers. The necessity of this assumption is clearly shown by Kleene's example of a tree well-founded with respect to recursive sequences but refuting the bar theorem. (Cf. Kleene and Vesley (1965), Lemma 9.8, bearing in mind that the fan theorem is a corollary of the bar theorem). As we shall see $\mathrm{Cont}_{\mathrm{LS}}$ = Ind conversely implies the bar theorem. See also the discussion in Dummett (1976).

2.13. *Historical remark; Brouwer's 'proof' of the bar theorem.*

In Brouwer (1927), there is a footnote (footnote 7 on page 63) to the effect that BI is seen to be true by reflection (i.e. no reduction to simpler insights is deemed necessary by Brouwer). However, Brouwer has given a much more elaborate argument for (a statement equivalent to) BI, which has received much attention, and concerning which Brouwer states in the above mentioned footnote (translated from the German original): 'nevertheless, the proof given in the text for the latter property seems to me to be of interest because of the assertions made in the course of it'. Stated informally, Brouwer's argument can be summarized as follows: how can we be certain that $\forall\alpha\exists x P(\overline{\alpha}x)$? If we define $P^{*}n$ as $\forall\alpha\in n\exists x P(\overline{\alpha}x)$, the question becomes: how can one be certain of $P^{*}0$? Brouwer's answer is: a 'fully analysed' proof of $P^{*}0$ must necessarily have a structure which can be

---

[†]That is to say, we cannot assume that, for each sequence in the range of α, some non-trivial intensional information will become available at some stage, so a method assigning an x to each α should not depend on such information.

described as a well-founded tree of inferences, built up from three types of inference:

η-inference	$\dfrac{Pn}{P^*n}$	(at terminal nodes)
ζ-inference	$\dfrac{P^*n}{P^*(n*\hat{x})}$	(for numerals n, x)
θ-inference	$\dfrac{P^*(n*\hat{0}), P^*(n*\hat{1}), P^*(n*\hat{2}), \ldots}{P^*n}$	

(for a numeral n; θ-inferences have infinitely many premises). One can then show, using well-foundedness of the tree structure that ζ-inferences can in fact be eliminated. The argument is then completed as follows: assume the premise of BI to be given. Proceeding up the tree, we can establish, parallel to the structure of η- and θ-inferences, Qn at each node where P^*n occurs (letting $Pn \to Qn$ and $\forall y Q(n*\hat{y}) \to Qn$ correspond to η- and θ-inferences respectively).

It is not the latter part of the argument which is in doubt; this is nothing else than induction over an inductively defined well-founded tree. But the assumption that to every proof we can associate a 'fully analysed proof' is problematic. It corresponds to cut-elimination or normalization for proofs of Π^1_1-statements; and here it is relevant to note that the bar theorem or equivalently 'Cont = Ind' is *classically* valid (for a proof see e.g. Kreisel and Troelstra (1970), Appendix, or Kleene and Vesley (1965), section 6.7 or Dummett (1976)); the proof of this fact runs parallel to the proof that classical logic with cut-free ω-rule is *complete* for Π^1_1-statements. However, the theory resulting by the addition of the bar theorem is certainly 'coherent' (cf. section 3.15 below); that is to say, the resulting theory is not only consistent, but the assumption implicit in the bar theorem is a 'natural' strong interpretation of the meaning of $\forall\alpha\exists x P\bar{\alpha}x$.

2.14. *The properties of* 'Ind' *expressed by neighbourhood functions; derivation of* $\mathrm{BI_D}$.

Corresponding to the inductive definition of 'Ind',

the neighbourhood functions representing elements of Ind may be introduced induvtively as follows

K1 $\lambda n.y+1 \in K$

K2 $a0=0 \ \& \ \forall x(\lambda n.a(\hat{x}*n)\in K) \to a \in K$,

and if we put

$$A(a,Q) \equiv a = \lambda n.y+1 \vee [a0=0 \ \& \ \forall x(\lambda n.a(\hat{x}*n)\in Q)],$$

then K1, K2 may be expressed as

$$A(a,K) \to Ka,$$

and the principle of induction over K (corresponding to induction over Ind) becomes

K3 $\forall a[A(a,Q) \to Qa] \to \forall a[Ka \to Qa]$.

Equation (3) of section 2.12 can now be formulated as

$$K_{LS} = K.$$

Now we can also show that the identification of continuous and inductively defined functionals implies the bar theorem. We first establish

Theorem. (Induction over unsecured sequences.)

$$\forall a\in K(\forall n(an \neq 0 \to Qn) \ \& \ \forall n(\forall yQ(n*\hat{y}) \to Qn)) \to Q0).$$

Proof: Apply induction over K (i.e. use the induction schema K3) with respect to the following predicate Aa for Qa in K3

$$Aa \equiv \forall m[\forall n(an \neq 0 \to Q(m*n)) \ \& \ \& \ \forall n(\forall yQ(m*n*\hat{y}) \to Q(m*n)) \to Qm].$$

Theorem. From $K_{LS} = K$ we can prove BI_D.

Proof: Let

(a) $\forall\alpha\exists xP\bar{\alpha}x$

(b) $\forall n(Pn \to Qn)$

(c) $\forall n(Pn \vee \neg Pn)$

(d) $\forall n(\forall yQ(n*\hat{y}) \to Qn)$.

Since P is decidable, the function e defined by

$$\begin{cases} en = 1 \text{ iff } \exists m\prec n(Pm) \\ en = 0 \text{ otherwise} \end{cases}$$

is an element of K_{LS} by virtue of (a).

Now consider $Q'n \equiv_{def} Qn \vee \exists m\prec n(Pm)$, and assume $\forall yQ'(n*\hat{y})$. If $\exists m\prec n(Pm)$ or Pn, then $Q'n$; so assume $\forall m\preceq n\neg(Pm)$, i.e.

$\neg\exists m{<}n*\hat{y}(Pm)$ for all y, then $\forall y Q'(n*\hat{y})$ implies $\forall y Q(n*\hat{y})$, hence Qn (by (d)), so once again $Q'n$. As a result

$$\forall y Q'(n*\hat{y}) \rightarrow Q'n.$$

It is also obvious that

$$en \neq 0 \rightarrow Q'n$$

and thus by induction over unsecured sequences, $K_{\mathrm{LS}} = K$ implies $Q'0$, hence $Q0$ (since $\exists m{<}0(Pm)$ is impossible). This establishes BI_{D}.

2.15. *Continuity axioms: final formulation.*

Let us now use $e, e', e'', f, f', f'', \ldots$ as variables ranging over K, and let us use the notation $e(\xi)$ (ξ any sort of functor) with definition

$$e(\xi)=y \equiv_{\mathrm{def}} \exists x(e(\overline{\xi}x) = y+1).$$

Then the strongest form of $\forall\alpha\exists x$-continuity, implicit in the assumption $K_{\mathrm{LS}} = K$, can be expressed as

$$\forall\alpha\exists x A(\alpha,x) \rightarrow \exists e\forall\alpha A(\alpha,e(\alpha)),$$

which is equivalent to

$$\forall\alpha\exists x A(\alpha,x) \rightarrow \exists e\forall n(en\neq 0 \rightarrow \forall\alpha\in n A(\alpha,en\dot{-}1)),$$

and similarly

$$\forall\alpha\exists a A(\alpha,a) \rightarrow \exists e\exists b\forall n(en\neq 0 \rightarrow \forall\alpha\in n A(\alpha,(b)_{en\dot{-}1})).$$

We have not yet reached full generality - so far we have not considered the case where A contains other lawless parameters besides α. Formal analogy with the case of LS 3 might tempt us to try

(1) $\quad \underline{\forall}\alpha\exists x A(\alpha,\beta_1,\ldots,\beta_n,x) \rightarrow \exists e\underline{\forall}\alpha A(\alpha,\beta_1,\ldots,\beta_n,e(\alpha)$,

but this is easily seen to be false, since the continuous operator required by $\underline{\forall}\alpha\exists x$ may well depend on $\beta_1,\beta_2,\ldots,$ as is seen by the following example. $\forall\alpha\forall\beta\exists x(\alpha 0+\beta 0=x)$ is obviously true; apply (1) to $\underline{\forall}\beta\exists x(\alpha 0+\beta 0=x)$ and we find as a result $\exists e\underline{\forall}\beta(\alpha 0+\beta 0=e(\beta))$. This is obviously false since e should essentially depend on α and therefore cannot be lawlike.

Instead of adopting (1), we may argue as follows. If $\alpha_1,\ldots,\alpha_p$ is a finite set of independent lawless sequences, i.e. $\#(\alpha_1,\ldots,\alpha_p)$, then at any stage we only know initial segments of $\alpha_1,\ldots,\alpha_p$; so an operation acting on $\alpha_1,\ldots,\alpha_p$ must be computed using such initial segments only. If we put

$$\nu_p(\alpha_1,\ldots,\alpha_p) \equiv_{\text{def}} \lambda x.\nu_p(\alpha_1 x,\ldots,\alpha_p x),$$

then an initial segment of length u of $\nu_p(\alpha_1,\ldots,\alpha_p)$ codes p initial segments of the same length u of $\alpha_1,\ldots,\alpha_p$, and operations on $\alpha_1,\ldots,\alpha_p$ may always be supposed to be computed from initial segments of $\nu_p(\alpha_1,\ldots,\alpha_p)$. So we can put

LS4 $$\underline{\forall}\alpha_1\ldots\underline{\forall}\alpha_p\exists aA(\alpha_1,\ldots,\alpha_p,a) \rightarrow \\ \rightarrow \exists e\exists b\underline{\forall}\alpha_1\ldots\underline{\forall}\alpha_p A(\alpha_1,\ldots,\alpha_p,(b)_{e(\nu_p(\alpha_1,\ldots,\alpha_p))}).$$

In other words, with respect to operations of type $N^N \rightarrow N$, p-tuples of independent lawless sequences behave like single lawless sequences. Of course, LS4 implies as a special case

$$\underline{\forall}\alpha_1\ldots\underline{\forall}\alpha_p\exists xA(\alpha_1,\ldots,\alpha_p,x) \rightarrow \\ \rightarrow \exists e\underline{\forall}\alpha_1\ldots\underline{\forall}\alpha_p A(\alpha_1,\ldots,\alpha_p,e(\nu_p(\alpha_1,\ldots,\alpha_p))).$$

2.16. Remark on the contrast between proof-theoretic strength and mathematical strength.

There is, in general, no direct correlation between proof-theoretic strength (as measured by the possibilities of proving the consistency of a system $\underset{\sim}{S}$ in a system $\underset{\sim}{S}'$) and mathematical strength (as measured by interesting or unexpected mathematical consequences and, above all, simpler proofs). So, in our case, the mathematically interesting and new (as compared with classical) results are all consequences of weak continuity schemata not involving the identification of continuous with inductively defined functionals, or, occasionally, of the fan theorem (stated in section 2.18); but, as illustrated by results in Chapter 5, these schemata do not increase proof-theoretic strength beyond first order arithmetic, whereas the addition of the bar theorem does not result in really striking new mathematical theorems but causes a great increase in proof-theoretic strength (see e.g. Howard and Kreisel 1966). It should also be noted that once K1-3 are accepted, which from an intuitive point of view are perfectly straightforward, we have already all the proof-theoretic strength of the bar theorem; for the addition of LS4, although requiring a new insight (i.e. the addition of LS4 is not intuitively straightforward) does not result in additional proof-theoretic strength, as will be seen from the elimination results in the next section.

2.17. Remark on the validity of LS 1-4.

We have established LS 1-4 as principles valid for lawless sequences; we did not describe a specific language to which the predicates appearing in the formulation of LS3, LS4 should be restricted.

The principles LS3, LS4 are obviously valid if the predicates considered do depend on lawless sequences as the only type of non-lawlike parameters, and are extensional with respect to lawlike parameters (our assumptions throughout the discussion). It is not necessary to require extensionality for lawless arguments explicitly, since $\alpha \equiv \beta \leftrightarrow \alpha = \beta$.

LS3 seems also to be justified for predicates containing, besides lawlike and lawless parameters, parameters for non-lawlike objects not constructed from lawless sequences (or constructed from a second universe of lawless sequences disjoint from the range of the lawless parameters), provided extensionality holds with respect to all function and set parameters.

In the case of LS4, the presence of parameters for non-lawlike objects (not lawless sequences) requires e to depend on them, hence e is no longer *lawlike* in the presence of such parameters.

2.18. Lawless sequences belonging to a fan.

Let a be a lawlike function; we say that α is bounded by a ($\alpha \leq a$) if $\forall x(\alpha x \leq ax)$. A sequence is lawless relative to the bounding function a if at any stage only an initial segment is known, together with the general condition $\forall x(\alpha x \leq ax)$. LS 1-3 can then be justified as before (with the obvious modification in LS1 and LS3 that the initial segments n must satisfy $\forall x < \mathrm{lth}(n)((n)_x \le ax))$, and with LS4 replaced by the fan theorem

$$\text{FAN*} \quad \underline{\forall}\alpha_1 \ldots \underline{\forall}\alpha_p \exists z A(\alpha_1,\ldots,\alpha_p,z) \to \\ \to \exists z \underline{\forall}\alpha_1 \ldots \underline{\forall}\alpha_p \exists y \underline{\forall}\beta_1 \ldots \underline{\forall}\beta_p (\overline{\beta}_1 z = \overline{\alpha}_1 z \ \& \ldots \& \overline{\beta}_p z = \overline{\alpha}_p z \\ \to A(\beta_1,\ldots,\beta_p,y))$$

where $\alpha_1,\ldots,\alpha_p$, $\beta_1,\ldots,\beta_p$ are supposed to range over lawless sequences bounded by a fixed lawlike sequence. In this case FAN* can also be retained in the presence of certain non-lawlike parameters for objects not constructed from lawless sequences. The reason for this is that even if the inductively defined neighbourhood functions now depend on certain non-lawlike parameters, one can obtain for inductively defined neighbourhood functions uniformity on bounded subsets of N^N just as before, which then with the continuity axioms yields the fan theorem; and in the fan theorem the neighbourhood functions do not appear explicitly anymore.

To see the truth of this, note that the inductive definition of K is *relative* to the range of the variables a, b, c, d; and if non-lawlike parameters are introduced, reasoning parallel to sections 2.9 - 2.15 yields a K^* which is inductively defined relative to sequences depending on these parameters (i.e. the sequences are 'lawlike in' or 'lawlike relative to' certain non-lawlike parameters).

3
METAMATHEMATICS OF LAWLESS SEQUENCES

3.1. *Description of* $\underset{\sim}{\mathrm{IDB}}_1$.

We now describe a theory $\underset{\sim}{\mathrm{IDB}}_1$ 'Inductively Defined Brouwer-operations' obtained by adding to $\underset{\sim}{\mathrm{EL}}_1$ (elementary analysis, described in section 1.8) the axioms for the class K of inductively defined neighbourhood functions, i.e. K1-3 mentioned before. To be precise, we extend the language $\underset{\sim}{\mathrm{EL}}_1$ by addition of variables e, e', e'', f, f', f'' for elements of K, and introduction of additional rules of term formation:

(i) When ϕ is a K-functor, ϕ' a functor, then $\phi(\phi')$ is a numerical term, $\phi|\phi'$ a functor;

(ii) When ϕ is a K-functor, t a (numerical) term, then ϕt is a term;

(iii) When ϕ is a K-functor, t a term, then $\lambda' n.\phi(\langle t\rangle * n)$ and $\lambda' n.St$ are K-functors (n not free in t);

(iv) When ϕ,ϕ' are K-functors, then $\phi{:}\phi'$, $\phi;\phi'$, ϕ/ϕ', $\phi//\phi'$ are K-functors;

(v) For a certain primitive recursive function k $\lambda' m.k(n,m)$ is a K-functor. (This again corresponds to a provable closure condition on K.)

:, ;, /, // are certain primitive recursive operations on functions which map pairs of K-functions onto K-functions. For a global understanding of the results, their precise definition does not concern us; it is sufficient to know that they correspond to certain operations under which K is provably closed. For a precise definition see Kreisel and Troelstra (1970), 3.2.8, 3.2.11. Intuitively, :, ; correspond to functional composition, since $(e{:}f)|a = e|(f|a)$, $(e;f)(a) = e(f|a)$. Clause (v) corresponds to another provable closure condition on K, see Kreisel and Troelstra (1970) 3.2.6. Intuitively, k represents a functional which maps each function onto a function with initial segment n, leaving other values unchanged, i.e.

$$\begin{cases}(\lambda' m.k(n,m)|a)(x) = (n)_x & \text{for } x < \mathrm{lth}(n),\\ (\lambda' m.k(n,m)|a)(x) = ax & \text{for } x \geq \mathrm{lth}(n).\end{cases}$$

The axioms and schemata of $\underset{\sim}{\mathrm{EL}}_1$ are now extended to the new language, quantifier rules and axioms for K-functions are added, and we also add

F1 $\quad e(a) = x \leftrightarrow \exists y(e(\overline{a}y) = x+1)$

F2 $\quad (e|a)(x) = y \leftrightarrow \exists z(e(\hat{x}*\overline{a}z) = y+1)$

K1-3 $\quad \begin{cases} a0 = 0 \;\&\; \forall x \exists e(e=\lambda n.a(\hat{x}*n)) \to \exists e(e=a) \\ \text{K3 (with } K\phi \text{ replaced by } \exists e(e=\phi)) \end{cases}$

K-CON $\quad (\lambda' n.\phi[n])t = \phi[t].$

Remarks. (i) The reason for the distinction between λ' and λ is that we need to distinguish, syntactically, between K-functors and functors; e.g. $\lambda n.St$ is a functor, but $\lambda' n.St$ is a K-functor (extensionally equal to $\lambda n.St$). This emphasizes that $\lambda' n.St$ is *given* to us as an element of K.

(ii) The notation $\phi(\phi')$ for ϕ' a functor, ϕ a K-functor corresponds to the notation $e(\xi)=x$ in section 2.15; cf. axiom F1. $\phi|\phi'$ is in fact fully explained by axiom F2, which shows that '|' corresponds to functional application if ϕ is regarded as a (code for a) functional of type $N^N \to N^N$ (if ϕ codes Φ of type $N^N \to N^N$, then $\lambda' n.\phi(\langle x\rangle * n)$ codes $\lambda\alpha.[(\Phi\alpha)x]$ of type $N^N \to N$ for all x).

(iii) K1 is implicit in clause (iii) of term-formation rules.

Remark on the choice of primitives. Inspection of the proof of the elimination theorems, to be discussed in sections 3.3-15, shows that the proof depends not on the schema K3 in general, but only on certain closure conditions on K.

The principal closure conditions are given by K1, K2, closure under functional composition (i.e. the existence of ; with property $e(f|a) = (e;f)(a)$) and closure under substitution (i.e. the existence of $\wedge$ such that $(e|a)(f(a)) = (e\wedge f)(a)$). For $e \wedge f$ one may take $\lambda n.sg(fn).e(\langle fn \dot{-} 1\rangle * n)$; / is obtained as a special case of $\wedge$ (cf. Kreisel and Troelstra (1970) 3.2.2 (ix), 3.2.8, 7.3.2). Moreover, another closure condition is indicated in lemma 3.8.

It would be of interest to find an elegant and more or less 'minimal' set of closure conditions on K which would suffice for the elimination results of $\underset{\sim}{\mathrm{LS}}$ and $\underset{\sim}{\mathrm{CS}}$ (discussed in Chapter 5; cf. also Troelstra (1974) where a list of closure conditions is given which is sufficient for the elimination results; this list is certainly not optimal).

3.2. *Description of* $\underset{\sim}{\mathrm{LS}}$

We now extend $\underset{\sim}{\mathrm{IDB}}_1$ by addition of variables for lawless sequences α, β, γ, δ together with an extension of the rules of term formation:

(vi) If $\alpha_1,\ldots,\alpha_u$ are lawless variables, t a term, then $(\nu_u(\alpha_1,\ldots,\alpha_u))t$ is a term;

(vii) If ϕ is a K-functor, t a term, $\alpha_1,\ldots,\alpha_u$ lawless variables, then $\phi(\nu_u(\alpha_1,\ldots,\alpha_u))$, $(\phi|\nu_u(\alpha_1,\ldots,\alpha_u))t$ are terms.

Now the axioms F1-2 of $\underset{\sim}{\mathrm{IDB}}_1$ are extended to the case where a is replaced by $\nu_u(\alpha_1,\ldots,\alpha_u)$, and we add the schemata LS1-4 relative to this language. We call the resulting system $\underset{\sim}{\mathrm{LS}}$.

Convention. We introduce the abbreviations

$$e(\xi_1,\ldots,\xi_u) \text{ for } e(\nu_u(\xi_1,\ldots,\xi_u)),$$
$$e|(\xi_1,\ldots,\xi_u) \text{ for } e|\nu_u(\xi_1,\ldots,\xi_u).$$

3.3. *The elimination theorem.*

The following theorem is stated, with a sketch of the proof, in Kreisel (1968):

Theorem (elimination theorem). There exists a mapping τ of formulae of $\underset{\sim}{\mathrm{LS}}$ without free lawless variabes onto the formulae of $\underset{\sim}{\mathrm{IDB}}_1$ such that

(1) $\tau(A) \equiv A$ for A a formula of $\underset{\sim}{\mathrm{IDB}}_1$

(2) $\underset{\sim}{\mathrm{LS}} \vdash A \leftrightarrow \tau(A)$

(3) $\underset{\sim}{\mathrm{LS}} \vdash A \Leftrightarrow \underset{\sim}{\mathrm{IDB}}_1 \vdash \tau(A)$ (finitistically, i.e. provable in primitive recursive arithmetic).

(2) implies that $\tau(A)$ is a contextual definition of the meaning of quantification over lawless sequences, a definition in terms of the notions of natural number, (constructive

function, and K-function which is used to justify $\underset{\sim}{IDB}_1$. In fact, τ presents an explanation of quantification over lawless sequences (relative to the language of $\underset{\sim}{LS}$) for any model of $\underset{\sim}{IDB}_1$. This leaves open the possibility that LS1-4 might contain, expressed *via* lawless sequences, some insights concerning the objects of $\underset{\sim}{IDB}_1$ not expressed in $\underset{\sim}{IDB}_1$ itself; but this possibility is ruled out by (3). (1)-(3) together imply in fact that $\underset{\sim}{LS}$ is a conservative extension of $\underset{\sim}{IDB}_1$. To formulate the result in another way: the axioms LS1-4 obtained by the conceptual analysis of Chapter 2 completely characterize quentification over lawless sequences (relative to the language of $\underset{\sim}{LS}$). Thus, once the elimination theorem has been proved, it becomes possible to regard quantification over lawless sequences as a 'figure of speech'.

We should mention here another 'elimination' method for lawless sequences, which is sketched in some detail in Dragalin (1973); according to Dragalin, the method is inspired by the definition of validity in Beth models (for Beth models, see Chapter 7).

3.4.

Since the existing description in Kreisel (1968) is sketchy, we shall describe τ in detail in sections 3.13, 3.14 and outline a proof of the elimination theorem. With respect to technical details, especially for $\underset{\sim}{IDB}_1$, we shall have to refer frequently to Kreisel and Troelstra (1970), where a similar elimination theorem is proved for the theory $\underset{\sim}{CS}$ to be discussed in Chapter 5. As a first step we change the basic language of $\underset{\sim}{LS}$, thereby obtaining an equivalent system $\underset{\sim}{LS}^*$.

3.5. Description of $\underset{\sim}{LS}^*$.

In $\underset{\sim}{LS}^*$, our logical operators are &, $\vee$, $\rightarrow$ ($\neg$ is regarded as defined), $\forall x$, $\exists x$, $\forall e$, $\exists e$, $\forall a$, $\exists a$, $\underline{\forall}\alpha\epsilon t$, $\underline{\exists}\alpha\epsilon t$ (t not containing lawless variables, with the intuitive meaning as in section 2.7). $\underline{\forall}\alpha$, $\underline{\exists}\alpha$ may be regarded as abbreviations of $\underline{\forall}\alpha\epsilon 0$, $\underline{\exists}\alpha\epsilon 0$ respectively.

We describe $\underset{\sim}{LS}^*$ as a system of formulae closed with respect to lawless variables. The quantifier rules and axioms for $\forall x, \exists x$, $\forall e$, $\exists e$, $\forall a$, $\exists a$ are as usual, and for $\underline{\forall}\alpha\in t$, $\underline{\forall}\alpha\in t$ they are

$$\underline{\forall}\alpha_1\in t_1\ldots\underline{\forall}\alpha_p\in t_p\underline{\forall}\beta\in t(P(\alpha_1,\ldots,\alpha_p) \to Q(\alpha_1,\ldots,\alpha_p,\beta)) \Rightarrow$$
$$\Rightarrow \underline{\forall}\alpha_1\in t_1\ldots\underline{\forall}\alpha_p\in t_p(P(\alpha_1,\ldots,\alpha_p) \to \underline{\forall}\beta\in tQ(\alpha_1,\ldots,\alpha_p,\beta)),$$

$$\underline{\forall}\alpha_1\in t_1\ldots\underline{\forall}\alpha_p\in t_p\underline{\exists}\beta\in t(\underline{\forall}\gamma\in tP(\alpha_1,\ldots,\alpha_p,\gamma) \to P(\alpha_1,\ldots,\alpha_p,\beta)).$$

$$\underline{\forall}\alpha_1\in t\ \ldots\underline{\forall}\alpha_p\in t_p\underline{\forall}\beta\in t(Q(\alpha_1,\ldots,\alpha_p,\beta) \to P(\alpha_1,\ldots,\alpha_p)) \Rightarrow$$
$$\Rightarrow \underline{\forall}\alpha_1\in t_1\ldots\underline{\forall}\alpha_p\in t_p(\underline{\exists}\beta\in tQ(\alpha_1,\ldots,\alpha_p,\beta) \to P(\alpha_1,\ldots,\alpha_p)),$$

$$\underline{\forall}\alpha_1\in t_1\ldots\underline{\forall}\alpha_p\in t_p\underline{\forall}\beta\in t(P(\alpha_1,\ldots,\alpha_p,\beta) \to \underline{\exists}\gamma\in tP(\alpha_1,\ldots,\alpha_p,\gamma)),$$

$$\underline{\forall}\alpha_1\ldots\underline{\forall}\alpha_n\underline{\forall}\alpha\in tA(\alpha_1,\ldots,\alpha_n,\alpha) \leftrightarrow \underline{\forall}\alpha_1\ldots\underline{\forall}\alpha_n\underline{\forall}\alpha(\alpha\in t\to A(\alpha_1,\ldots,\alpha_n,\alpha)),$$

$$\underline{\forall}\alpha_1\ \ldots\underline{\forall}\alpha_n\underline{\exists}\alpha\in tA(\alpha_1,\ldots,\alpha_n,\alpha) \leftrightarrow \underline{\forall}\alpha_1\ldots\underline{\forall}\alpha_n\underline{\exists}\alpha(\alpha\in t\& A(\alpha_1,\ldots,\alpha_n,\alpha)).$$

Propositional rules and axioms and quantifier-rules and axioms for $\forall x$, $\exists x$, $\forall a$, $\exists a$, $\forall e$, $\exists e$ are stated as the $\underline{\forall}$-closures of the usual rules and axioms, e.g. if $F_1,\ldots,F_n \Rightarrow F$ is an instance of one of the usual rules, then $F_1^*,\ldots,F_n^* \Rightarrow F^*$ is an instance of the corresponding rule in $\underset{\sim}{LS}^*$, where $F_1^*,\ldots,F_n^*$, F^* are the $\underline{\forall}$-closures of $F_1,\ldots,F_n,F$ respectively, and where $F_1,\ldots,F_n,F$ are assumed to be expressed in the language of $\underset{\sim}{LS}^*$.

The essential feature distinguishing $\underline{\forall}$ from $\forall$ (and implicitly $\underline{\exists}$ from $\exists$) is expressed by

$$\underline{\forall}\alpha_1\ldots\underline{\forall}\alpha_p(\#(\alpha_1,\ldots,\alpha_p)).$$

The axioms LS1-4 can be restated as follows.

LS1 $\quad\forall n\underline{\exists}\alpha(\alpha\in n)$,

LS2 $\quad$(implicit in our use of $\underline{\forall}\alpha$, $\underline{\exists}\alpha$),

LS3 $\quad\underline{\forall}\alpha_1\ldots\underline{\forall}\alpha_p(A(\alpha_1,\ldots,\alpha_p) \to \exists n_1\ldots n_p(\alpha_1\in n_1\ \&\ldots\&\ \alpha_p\in n_p\ \&$
$\quad\&\underline{\forall}\beta_1\in n_1\ldots\underline{\forall}\beta_p\in n_pA(\beta_1,\ldots,\beta_p)))$,

LS4 $\quad\underline{\forall}\alpha_1\ldots\underline{\forall}\alpha_p\exists aA(\alpha_1,\ldots,\alpha_p,a) \to$
$\quad\to \exists e\exists b\underline{\forall}\alpha_1\ldots\underline{\forall}\alpha_pA(\alpha_1,\ldots,\alpha_p,(b)_{e(\alpha_1,\ldots,\alpha_p)})$.

Let us give two examples of deductions in $\underset{\sim}{LS}^*$; the results will be used in the sequel.

3.6.

<u>Example 1.</u> Deduction of $\forall n(\underline{\forall}\alpha\in nA\alpha \to \underline{\forall}\alpha\in nB\alpha) \to \underline{\forall}\alpha(A\alpha \to B\alpha)$.

(1) $\forall n(\underline{\forall}\alpha\in nA\alpha \to \underline{\forall}\alpha\in nB\alpha)$ (assumption)

(2) $\underline{\forall}\alpha(A\alpha \to \exists m(\alpha\in m \ \& \ \underline{\forall}\alpha'\in mA\alpha'))$ (LS3)

(3) $\underline{\forall}\alpha((\alpha\in m \ \& \ \underline{\forall}\alpha'\in mA\alpha') \to \underline{\forall}\alpha'\in mA\alpha')$

($\underline{\forall}$-closure of an instance of $P \ \& \ Q \to Q$)

(4) $\underline{\forall}\alpha((\alpha\in m \ \& \ \underline{\forall}\alpha'\in mA\alpha') \to \underline{\forall}\alpha'\in mB\alpha')$

(from (1), (3) by $\underline{\forall}$-closure of transitivity of $\to$)

(5) $\underline{\forall}\alpha'\in mB\alpha' \to \underline{\forall}\alpha'(\alpha'\in m \to B\alpha')$ (axiom)

(6) $\underline{\forall}\alpha((\alpha\in m \ \& \ \underline{\forall}\alpha'\in mA\alpha') \to \underline{\forall}\alpha'(\alpha'\in m \to B\alpha'))$

((4),(5), transitivity of $\to$)

(7) $\underline{\forall}\alpha(\underline{\forall}\alpha'(\alpha'\in m \to B\alpha') \to (\alpha\in m \to B\alpha))$ (quantification axiom)

(8) $\underline{\forall}\alpha((\alpha\in m \ \& \ \underline{\forall}\alpha'\in mA\alpha') \to B\alpha)$ ((6),(7), transitivity of $\to$, modus ponens)

(9) $\underline{\forall}\alpha(A\alpha \to B\alpha)$ ((8),(2), transitivity of $\to$).

Finally (1) $\to$ (9).

3.7.

<u>Example 2.</u> Deduction of $\underline{\forall}\alpha(\underline{\exists}\beta\in t \ A(\alpha,\beta) \to \exists n\underline{\forall}\beta\in t{*}n \ A(\alpha,\beta))$.

(1) $\underline{\forall}\alpha\underline{\forall}\gamma\in t(\underline{\exists}\beta\in t \ A(\alpha,\beta) \to A(\alpha,\gamma))$ (quantifier axiom)

(2) $\underline{\forall}\alpha\underline{\forall}\beta[\beta\in t \ \& \ A(\alpha,\beta) \to$

$\to \exists nm[\alpha\in n \ \& \ \beta\in m \ \& \ \underline{\forall}\alpha'\in n\underline{\forall}\beta'\in m(\beta'\in t \ \& \ A(\alpha',\beta'))]]$

(application of LS3)

(3) $\underline{\forall}\alpha\underline{\forall}\beta[\beta\in t \to (A(\alpha,\beta) \to$

$\to \exists nm(\alpha\in n \ \& \ \underline{\forall}\alpha'\in n\underline{\forall}\beta'\in m(\beta'\in t \ \& \ A(\alpha',\beta'))))]$

(propositional logic, from (2))

(4) $\underline{\forall}\alpha\underline{\forall}\beta\in t[A(\alpha,\beta) \to \exists nm(\alpha\in n \ \& \ \underline{\forall}\alpha'\in n\underline{\forall}\beta'\in m(\beta'\in t \ \& \ A(\alpha',\beta')))]$

(axiom)

(5) $\underline{\forall}\alpha(\underline{\exists}\beta\in t \ A(\alpha,\beta) \to \exists nm[\alpha\in n \ \& \ \underline{\forall}\alpha'\in n\underline{\forall}\beta'\in m(\beta'\in t \ \& \ A(\alpha',\beta'))])$

(from (4) by a quantifier rule)

(6) $\underline{\forall}\alpha(\underline{\exists}\beta\in t \ A(\alpha,\beta) \to \exists m\underline{\forall}\beta'\in m(\beta'\in t \ \& \ A(\alpha,\beta')))$

(by propositional logic from (5))

(7) $\underline{\forall}\alpha(\underline{\exists}\beta\in t \ A(\alpha,\beta) \to \exists n\underline{\forall}\beta'\in t{*}n(\beta'\in t \ \& \ A(\alpha,\beta')))$

(by propositional logic and arithmetic from (6); one must show that $\underline{\forall}\beta'\in m(\beta'\in t)$ implies $m \succeq t$).

(8) $\underline{\forall}\alpha(\underline{\exists}\beta\in t \ A(\alpha,\beta) \to \exists n\underline{\forall}\beta\in t{*}n \ A(\alpha,\beta'))$

(propositional logic).

In order to prove the first half (i.e. (1),(2)) of the elimination theorem, we need a number of formal consequences of these axioms (collected together in theorem 3.12) for which in turn we need some lemmas.

3.8.

Lemma. If we define, for each K-function f, and any p-tuple $m_1,\ldots,m_p$ of natural numbers a function $f[m_1,\ldots,m_k]$ by

$$(f[m_1,\ldots,m_k])m = y+1 \leftrightarrow$$
$$\exists m'(fm'=y+1\&[(k^p m' \preceq m_1 * k_1^p m)\&\ldots\&(k_p^p m' \preceq m_p * k_p^p m)]),$$

then $f[m_1,\ldots,m_k] \in K$.

Proof: Apply induction over K with respect to f to establish $\forall m_1\ldots m_p(f[m_1,\ldots,m_p] \in K)$.

3.9.

Lemma. $\underline{\forall}\alpha \ldots \underline{\forall}\alpha_p \exists x(e(\nu_p(\overline{\alpha_1,\ldots,\alpha_p})x) \neq 0)$.

Proof: We establish the lemma by proving a stronger asser tion. By induction over K with respect to e we establish

$$(1) \qquad \underline{\forall}\alpha_1\ldots\underline{\forall}\alpha_p \forall x \exists yz(\nu_p(\alpha_1,\ldots,\alpha_p)\epsilon x \to$$
$$\to ey=Sz \;\&\; \nu_p(\alpha_1,\ldots,\alpha_p)\epsilon x*y).$$

For $e = \lambda'n.Su$, we may take $0,u$ for y,z respectively. Now suppose $e0 = 0$, $\text{lth}(x) = u$, $\nu_p(\alpha_1,\ldots,\alpha_p)(u) = u'$. The induction hypothesis tells us that

$$\forall v\underline{\forall}\alpha_1\ldots\underline{\forall}\alpha_p \forall x \exists yz(\nu_p(\alpha_1,\ldots,\alpha_p)\epsilon x \to$$
$$\to (\lambda'n.e(\hat{v}*n))(y) = Sz \;\&\; \nu_p(\alpha_1,\ldots,\alpha_p)\epsilon x*y).$$

Apply this with u', $x*\langle u'\rangle$ for v, x, then

$\nu_p(\alpha_1,\ldots,\alpha_p)\epsilon x*\langle u'\rangle \to e(\langle u'\rangle*y) = Sz \;\&\; \nu_p(\alpha_1,\ldots,\alpha_p)\epsilon x*\langle u'\rangle*y$ for suitable y, z, and hence we may take $\langle u'\rangle*y$, z for y,z in (1).

Remark. This proof (from Kreisel (1968)) does not assume any closure properties on sequences of the form $\nu_p(\alpha_1,\ldots,\alpha_p)$; the only essential property is that to each argument a value is determined. So quite generally, for any sort of sequence ξ

$$\exists x(e(\overline{\xi}x) \neq 0).$$

3.10.

Lemma. $\forall n \forall m(en \neq 0 \rightarrow e(n*m) = en)$.

Proof: as in Kreisel and Troelstra (1970), 3.2.10 (ii), by induction with respect to K.

3.11.

Lemma. For any $t[\alpha_1,\ldots,\alpha_p]$ of $\underset{\sim}{LS}*$, we can (in a suitable conservative extension of $\underset{\sim}{LS}*$) find a K-functor e_t such that

$$\begin{cases} \underset{\sim}{IDB}_1 \vdash e_t(a_1,\ldots,a_p) = t[a_1,\ldots,a_p], \\ \underset{\sim}{LS}* \vdash e_t(\alpha_1,\ldots,\alpha_p) = t[\alpha_1,\ldots,\alpha_p]. \end{cases}$$

Proof: By induction on the construction of t, very similar to the proof of 7.3.2 in Kreisel and Troelstra (1970). Let us consider two cases for illustration.

(a) Let e_t, $e_{t'}$ for $t[\alpha_1,\alpha_2,x]$ and $t'[\alpha_1,\alpha_2]$ be known already. We wish to construct the corresponding $e_{t''}$ for $t'' \equiv t[\alpha_1,\alpha_2,t'[\alpha_1,\alpha_2]]$. e_t is a functor $\phi[x]$; let e_t be defined such that

$$\tilde{e}_t 0 = 0,\ \tilde{e}_t(\hat{x}*n) = \phi[x](n).$$

Then take for $e_{t''}$

$$e_{t''} \equiv \lambda' n.\tilde{e}_t.\mathrm{sg}(e_{t'}n).(\langle e_{t'}n \dot{-} 1\rangle * n).$$

By Kreisel and Troelstra (1970), 3.2.2 (ix) this is indeed a K-function.

(b) Assume t to be defined by primitive recursion from t', t'', i.e.

$$t[\alpha_1,\alpha_2,x] = Rt't''x, \quad \text{i.e.}$$

$$\begin{cases} t[\alpha_1,\alpha_2,0] = t'[\alpha_1,\alpha_2], \\ t[\alpha_1,\alpha_2,Sz] = t''[j(t[\alpha_1,\alpha_2,z],z),\alpha_1,\alpha_2], \end{cases}$$

and assume $e_{t'}$, $e_{t''}$ to be already constructed. Note also that the addition of a constant R_K to the language satisfying

$$R_K ef0 = 0,\ R_K ef(\hat{0}*n) = en,$$
$$R_K ef(\langle Sz\rangle * n) = f(\langle j(R_K ef(\hat{z}*n) \dot{-} 1, z)\rangle * n)$$

is a conservative extension, since it is readily proved by induction that $\forall x(\lambda n.R_K ef(\hat{x}*n) \in K)$.

Now take $e_t \equiv R_K e_{t'}\tilde{e}_{t''}$, where $\tilde{e}_{t''}$ has been chosen such that

$\tilde{e}_{t''}0 = 0,\ \tilde{e}_{t''}(\hat{x}*n) = \phi[x](n),$
and where $\phi[x]$ is the functor $e_{t''}$ representing $t''[x,\alpha_1,\alpha_2]$.

3.12.

Theorem.

(a) $\underline{\forall}\alpha_1\epsilon t_1 \ldots \underline{\forall}\alpha_p\epsilon t_p(t[\alpha_1,\ldots,\alpha_p] = t'[\alpha_1,\ldots,\alpha_p]) \leftrightarrow$
$\forall a_1\epsilon t_1\ldots\forall a_p\epsilon t_p(t[a_1,\ldots,a_p] = t'[a_1,\ldots,a_p])$

(b) $\underline{\exists}\beta\epsilon t\ A(\alpha_1,\ldots,\alpha\ ,\beta) \leftrightarrow \exists n\underline{\forall}\beta\epsilon t*n\ A(\alpha_1,\ldots,\alpha_p,\beta)$

(c) $\underline{\forall}\alpha_1\epsilon t_1 \ldots \underline{\forall}\alpha_p\epsilon t_p(A(\alpha_1,\ldots,\alpha_p) \to B(\alpha_1,\ldots,\alpha_p)) \leftrightarrow$
$\forall n_1\ldots n_p(\underline{\forall}\alpha_1\epsilon t_1*n_1\ldots\underline{\forall}\alpha_p\epsilon t_p*n_pA(\alpha_1,\ldots,\alpha_p) \to$
$\underline{\forall}\alpha_1\epsilon t_1*n_1\ldots\underline{\forall}\alpha_p\epsilon t_p*n_p\ B(\alpha_1,\ldots,\alpha_p))$

(d) $\underline{\forall}\alpha_1\epsilon t_1\ldots\underline{\forall}\alpha_p\epsilon t_p\exists xA(\alpha_1,\ldots,\alpha_p,x) \leftrightarrow$
$\exists e\forall n(en\neq 0\to\underline{\forall}\alpha_1\epsilon t_1*k_1^pn\ldots\underline{\forall}\alpha_p\epsilon t_p*k_p^pn\ A(\alpha_1,\ldots,\alpha_p,en\dot{-}1))$

(e) $\underline{\forall}\alpha_1\epsilon t_1\ldots\underline{\forall}\alpha_p\epsilon t_p\exists a\ A(\alpha_1,\ldots,\alpha_p,a) \leftrightarrow$
$\exists e\exists b\forall n(en\neq 0 \to$
$\underline{\forall}\alpha_1\epsilon t_1*k_1^pn\ldots\underline{\forall}\alpha_p\epsilon t_p*k_p^pnA(\alpha_1,\ldots,\alpha_p,\lambda m.b(\langle en\dot{-}1\rangle*m)))$

(f) $\underline{\forall}\alpha_1\epsilon t_1\ldots\underline{\forall}\alpha_p\epsilon t_p\exists e\ A(\alpha_1,\ldots,\alpha_p,e) \leftrightarrow$
$\exists e\exists f\forall n(en\neq 0 \to$
$\underline{\forall}\alpha_1\epsilon t_1*k_1^pn\ldots\underline{\forall}\alpha_p\epsilon t_p*k_p^pnA(\alpha_1,\ldots,\alpha_p,\lambda' m.f(\langle en\dot{-}1\rangle*m)))$.

Proof:

(a) Because of lemma 3.11, we only have to prove

$$\underline{\forall}\alpha_1\epsilon t_1\ldots\underline{\forall}\alpha_p\epsilon t_p(e(\alpha_1,\ldots,\alpha_p) = f(\alpha_1,\ldots,\alpha_p)) \leftrightarrow$$
$$\forall a_1\epsilon t_1\ldots\forall a_p\epsilon t_p(e(a_1,\ldots,a_p) = f(a_1,\ldots,a_p)).$$

Assume $\underline{\forall}\alpha_1\epsilon t_1\ldots\underline{\forall}\alpha_p\epsilon t_p(e(\alpha_1,\ldots,\alpha_p) = f(\alpha_1,\ldots,\alpha_p))$, and let $a_1\epsilon t_1,\ldots,a_p\epsilon t_p$. We can find an n such that $\nu_p(a_1,\ldots,a_p)\epsilon n$, $en\neq 0$, $fn\neq 0$ (e.g. lemma 3.9). Take $\alpha_1\epsilon t_1,\ldots,\alpha_p\epsilon t_p$, $\nu_p(\alpha_1,\ldots,\alpha_p)\epsilon n$, $\#(\alpha_1,\ldots,\alpha_p)$; then $e(\alpha_1,\ldots,\alpha_p) = f(\alpha_1,\ldots,\alpha_p) = en\dot{-}1$, hence also $e(a_1,\ldots,a_p) = f(a_1,\ldots,a_p)$; similarly in the other direction.

(b) See example 1 (section 3.6).

(c) See example 2 (section 3.7).

(d) We need lemma 3.8. From LS4 it follows that if we assume

$$\underline{\forall}\alpha_1\ldots\underline{\forall}\alpha_p\exists x(\alpha_1\epsilon t_1\ \&\ldots\&\ \alpha_p\epsilon t_p \to A(\alpha_1,\ldots,\alpha_p,x)),$$

then $\underline{\forall}\alpha_1\ldots\underline{\forall}\alpha_p(\alpha_1\epsilon t_1\ \&\ldots\&\ \alpha_p\epsilon t_p \to A(\alpha_1,\ldots,\alpha_p,e(\alpha_1,\ldots,\alpha_p)))$ for some e. Then $e[t_1,\ldots,t_p]$ satisfies

$\forall n(e[t_1,\ldots,t_p]n\neq 0 \rightarrow$
$\underline{\forall}\alpha_1\epsilon t_1{*}k_1^p n\ldots\underline{\forall}\alpha_p\epsilon t_p{*}k_p^p nA(\alpha_1,\ldots,\alpha_p,e[t_1,\ldots,t_p]n\dot{-}1)).$

(e),(f) Similarly.

3.13. Definition of τ *for* $\underset{\approx}{\mathrm{LS}}^*$.

Let p be the number of logical operators and prime formulae, but not counting $\underline{\forall}\alpha\epsilon t, \underline{\exists}\alpha\epsilon t$, occurring within the scope of a quantifier $\underline{\forall}\alpha\epsilon t, \underline{\exists}\alpha\epsilon t$ in A, and let q be the number of quantifiers $\underline{\exists}\alpha\epsilon t$ plus the number of disjunctions within the scope of $\underline{\forall}\alpha\epsilon t, \underline{\exists}\alpha\epsilon t$ in A. Then the *degree* of A is defined as the ordinal $\omega q + p$. We shall now define an auxiliary mapping $\mapsto$, defined for formulae closed with respect to lawless quantifiers, which lowers the degree of such formulae.

(i) $\underline{\forall}\alpha_1\epsilon t_1\ldots\underline{\forall}\alpha_p\epsilon t_p\ A(\alpha_1,\ldots,\alpha_p) \mapsto$
$\forall a_1\epsilon t_1\ldots\forall a_p\epsilon t_p A(a_1,\ldots,a_p)$ for A prime.

(ii) $\underline{\exists}\alpha\epsilon tA \mapsto \exists n\underline{\forall}\alpha\epsilon t{*}nA.$

(iii) $\underline{\forall}\alpha_1\epsilon t_1\ldots\underline{\forall}\alpha_p\epsilon t_p(A(\alpha_1,\ldots,\alpha_p) \rightarrow B(\alpha_1,\ldots,\alpha_p)) \mapsto$
$\forall n_1\ldots n_p(\underline{\forall}\alpha_1\epsilon t_1{*}n_1\ldots\underline{\forall}\alpha_p\epsilon t_p{*}n_pA(\alpha_1,\ldots,\alpha_p) \rightarrow$
$\rightarrow \underline{\forall}\alpha_1\epsilon t_1{*}n_1\ldots\underline{\forall}\alpha_p\epsilon t_p{*}n_pB(\alpha_1,\ldots,\alpha_p)).$

(iv) $\underline{\forall}\alpha_1\epsilon t_1\ldots\underline{\forall}\alpha_p\epsilon t_p(A(\alpha_1,\ldots,\alpha_p)\ \&\ B(\alpha_1,\ldots,\alpha_p)) \mapsto$
$\underline{\forall}\alpha_1\epsilon t_1\ldots\underline{\forall}\alpha_p\epsilon t_pA(\alpha_1,\ldots,\alpha_p)\&\underline{\forall}\alpha_1\epsilon t_1\ldots\underline{\forall}\alpha_p\epsilon t_pB(\alpha_1,\ldots,\alpha_p).$

(v) $\underline{\forall}\alpha_1\epsilon t_1\ldots\underline{\forall}\alpha_p\epsilon t_p(A\vee B) \mapsto$
$\underline{\forall}\alpha_1\epsilon t_1\ldots\underline{\forall}\alpha_p\epsilon t_p\exists x[(x=0\rightarrow A)\ \&\ (x\neq 0\rightarrow B)].$

(vi) $\underline{\forall}\alpha_1\epsilon t_1\ldots\underline{\forall}\alpha_p\epsilon t_p\exists xA(\alpha_1,\ldots,\alpha_p,x) \mapsto$
$\exists e\forall n(en\neq 0\rightarrow\underline{\forall}\alpha_1\epsilon t_1{*}k_1^p n\ldots\underline{\forall}\alpha_p\epsilon k_p^p nA(\alpha_1,\ldots,\alpha_p,en\dot{-}1)).$

(vii) $\underline{\forall}\alpha_1\epsilon t_1\ldots\underline{\forall}\alpha_p\epsilon t_p\exists aA(\alpha_1,\ldots,\alpha_p,a) \rightarrow \exists e\exists b\forall n(en\neq 0 \rightarrow$
$\underline{\forall}\alpha_1\epsilon t_1{*}k_1^p n\ldots\underline{\forall}\alpha_p\epsilon t_p{*}k_p^p nA(\alpha_1,\ldots\alpha_p,\lambda m.b(\langle en\dot{-}1\rangle{*}m))).$

(viii) $\underline{\forall}\alpha_1\epsilon t_1\ldots\underline{\forall}\alpha_p\epsilon t_p\exists eA(\alpha_1,\ldots,\alpha_p,e) \mapsto$
$\exists e\exists f\forall n(en\neq 0\rightarrow$
$\underline{\forall}\alpha_1\epsilon t_1{*}k_1^p n\ldots\underline{\forall}\alpha_p\epsilon t_p{*}k_p^p nA(\alpha_1,\ldots,\alpha_p,\lambda' m.f(\langle en\dot{-}1\rangle{*}m))).$

(ix) $\underline{\forall}\alpha_1\epsilon t_1\ldots\underline{\forall}\alpha_p\epsilon t_p\underline{\exists}\beta\epsilon tA(\alpha_1,\ldots,\alpha_p,\beta) \mapsto$
$\underline{\forall}\alpha_1\epsilon t_1\ldots\underline{\forall}\alpha_p\epsilon t_p\exists n\underline{\forall}\beta\epsilon t{*}nA(\alpha_1,\ldots,\alpha_p,\beta).$

(x) $\underline{\forall}\alpha_1\epsilon t_1\ldots\underline{\forall}\alpha_p\epsilon t_p\forall xA(\alpha_1,\ldots,\alpha_p,x) \rightarrow$
$\forall x\underline{\forall}\alpha_1\epsilon t_1\ldots\underline{\forall}\alpha_p\epsilon t_pA(\alpha_1,\ldots,\alpha_p,x)$ and similarly for
$\underline{\forall}\alpha_1\epsilon t_1\ldots\forall aB(\alpha_1,\ldots,a),\underline{\forall}\alpha_1\epsilon t_1\ldots\forall eB(\alpha_1,\ldots,e).$

Now τ is defined as the result of applying $\mapsto$ till a formula with degree 0 has been reached.

3.14.

Theorem. For the translation τ as defined in section 3.13, it follows that for any formula A of $\underset{\sim}{LS}^*$:

(a) $\tau(A) \equiv A$ for formulae of $\underset{\sim}{IDB}_1$,

(b) $\underset{\sim}{LS}^* \vdash \tau(A) \leftrightarrow A$.

Proof: (a) is immediate. (b) follows from theorem 3.12 by noting that at each application of the auxiliary mapping $\mapsto$, a formula is replaced by one that is equivalent to it relative to $\underset{\sim}{LS}^*$.

3.15.

We shall now indicate how to obtain the second half of the elimination theorem. Let us use $\ulcorner A \urcorner$ to indicate $\tau(A)$. Then we have to establish the following lemmas:

Lemma. Let $t[\alpha_1,\ldots,\alpha_p]$, $t'[\alpha_1,\ldots,\alpha_p]$ be two terms such that in $\underset{\sim}{IDB}_1$

$$\vdash \forall a_1\ldots a_p(t[a_1,\ldots,a_p] = t'[a_1,\ldots,a_p]).$$

Then also in IDB_1

$$\vdash \ulcorner \underline{\forall}\alpha_1 \epsilon t_1 \ldots \underline{\forall}\alpha_p \epsilon t_p A(\alpha_1,\ldots,\alpha_p,t)\urcorner \leftrightarrow \ulcorner \underline{\forall}\alpha_1 \epsilon t_1 \ldots \underline{\forall}\alpha_p \epsilon t_p A(\alpha_1,\ldots,\alpha_p,t')\urcorner.$$

Similarly for functors instead of terms.

Proof: By induction on the logical complexity of A.

Lemma. $\underset{\sim}{IDB}_1 \vdash \ulcorner \underline{\forall}\alpha_1 \epsilon n_1 \ldots \underline{\forall}\alpha_p \epsilon n_p A(\alpha_1,\ldots,\alpha_p)\urcorner \rightarrow \ulcorner \underline{\forall}\alpha_1 \epsilon n_1 * m_1 \ldots \underline{\forall}\alpha_p \epsilon n_p * m_p A(\alpha_1,\ldots,\alpha_p)\urcorner$.

Proof: By induction on the logical complexity of A; compare the similar proof of 7.3.5 in Kreisel and Troelstra (1970).

Lemma. $\underset{\sim}{IDB}_1 \vdash \ulcorner \underline{\forall}\alpha_1 \epsilon n_1 \ldots \underline{\forall}\alpha_p \epsilon n_p \exists x A(\alpha_1,\ldots,\alpha_p,x)\urcorner \leftrightarrow \ulcorner \exists e \forall n(en \neq 0 \rightarrow \underline{\forall}\alpha_1 \epsilon n_1 * k_1^p n \ldots \underline{\forall}\alpha_p \epsilon n_p * k_p^p n A(\alpha_1,\ldots,\alpha_p, en \dot{-} 1))\urcorner$.

Proof: By induction on the logical complexity of A; cf. the proof of 7.3.6 in Kreisel and Troelstra (1970).

Corollary. $\underset{\sim}{IDB}_1 \vdash \ulcorner \underline{\forall}\alpha_1 \epsilon n_1 \ldots \underline{\forall}\alpha_p \epsilon n_p A(\alpha_1,\ldots,\alpha_p)\urcorner \leftrightarrow \ulcorner \exists e \forall n(en \neq 0 \rightarrow \underline{\forall}\alpha_1 \epsilon n_1 * k_1^p n \ldots \underline{\forall}\alpha_p \epsilon n_p * k_p^p n A(\alpha_1,\ldots,\alpha_p))\urcorner$.

Theorem. For the mapping τ as defined in section 3.13 we can prove finitistically (e.g. in primitive recursive arithmetic)

$$\underset{\sim}{\mathrm{LS}}^* \vdash A \Leftrightarrow \underset{\sim}{\mathrm{IDB}}_1 \vdash \tau(A).$$

The proof proceeds by induction on the length of derivations, cf. again the proof of 7.3.8 in Kreisel and Troelstra (1970).

3.16. Extension to LSS *and* LSS*.

Now we shall indicate how to extend LS to a theory containing variables for relations (= species) of natural numbers, lawlike functions, elements of K, and lawless functions. The relations themselves are thought of as lawlike (complete) objects, i.e. not dependent on for example lawless parameters. Therefore, if we use $X_{n,m,p,q}$ as a variable for a relation with n numerical, m lawlike, p K-function, and q lawless arguments, the appropriate axiom schema is

LS5 $$\underline{\forall}\alpha_1 \ldots \underline{\forall}\alpha_p \exists X_{n,m,p,q} A(X_{n,m,p,q}) \to \exists Y_{n+1,m,p,q} \exists e \forall n(en \neq 0 \to \underline{\forall}\alpha_1 \in k_1^p n \ldots \underline{\forall}\alpha_p \in k_p^p n A([1/en \dot{-} 1] Y_{n+1,m,p,q}))$$

where $A([1/en\dot{-}1]Y_{n+1,m,p,q})$ indicates the formula obtained from $A(X)$ by replacing each occurrence of $X\mathfrak{A}$ ($\mathfrak{A}$ standing for a string of arguments) by $Y(en\dot{-}1,\mathfrak{A})$.

In addition, the extension LSS of LS with species variables should contain the quantifier rules and axioms with respect to the new sort of variables, and suitable comprehension schemata. We shall here assume LSS to contain full impredicative comprehension:

CA $$\exists X_{n,m,p,q} \forall x_1 \ldots \forall a_1 \ldots \forall e_1 \ldots \forall \alpha_1 \ldots [A(x_1,\ldots,a_1,\ldots,e_1,\ldots,\alpha_1,\ldots) \leftrightarrow X_{n,m,p,q}(x_1,\ldots,a_1,\ldots,e_1,\ldots,\alpha_1,\ldots)]$$

where A therefore may contain bound species variables. (This is not intended to indicate our commitment to full impredicative comprehension as an intuitionistic principle, but to emphasize that the possibility of elimination of lawless variables remains even under the assumption of very strong comprehension principles).

For a convenient definition of τ, we must switch to an equivalent system $\underset{\sim}{\text{LSS}}$* with a different set of primitives. In this case, we not only replace $\forall\alpha, \exists\alpha$ as primitive logical operators by $\underline{\forall}\alpha\epsilon t, \underline{\exists}\alpha\epsilon t$ (t not containing lawless parameters), but we also make a change with respect to species variables: we shall assume species variables to apply to independent lawless arguments only, i.e.

$$X_{n,m,p,q}(\mathfrak{A},\alpha_1,\ldots,\alpha_q) \to \#(\alpha_1,\ldots,\alpha_q)$$

($\mathfrak{A}$ abbreviating the string of non-lawless arguments), and we shall regard $X_{n,m,p,q}(\mathfrak{A},\alpha_1,\ldots,\alpha_q)$ as a well-formed formula only when all variables $\alpha_1,\ldots,\alpha_q$ are distinct.
Of course CA must now be reformulated as

CA* $\quad \exists X_{n,m,p,q}\forall\mathfrak{A}\underline{\forall}\alpha_1\ldots\underline{\forall}\alpha_p[A(\mathfrak{A},\alpha_1,\ldots,\alpha_p) \leftrightarrow X_{n,m,p,q}(\mathfrak{A},\alpha_1,\ldots,\alpha_p)].$

That the theory $\underset{\sim}{\text{LSS}}$* thus obtained is in fact equivalent to $\underset{\sim}{\text{LSS}}$ is a consequence of the decidability of equality between lawless sequences: each predicate in $\underset{\sim}{\text{LSS}}$ can be decomposed into a number of predicates applicable to independent lawless arguments only. Consider for example a ternary predicate $A(\alpha,\beta,\gamma)$; this can be decomposed as follows:

$$\begin{aligned} A(\alpha,\beta,\gamma) \leftrightarrow [(\#(\alpha,\beta,\gamma) \,\&\, A(\alpha,\beta,\gamma)) \vee \\ \vee (\alpha=\beta \,\&\, \alpha\neq\gamma \,\&\, A(\alpha,\beta,\gamma)) \vee (\alpha=\gamma \,\&\, \alpha\neq\beta \,\&\, A(\alpha,\beta,\gamma)) \vee \\ \vee (\alpha\neq\beta \,\&\, \beta=\gamma \,\&\, A(\alpha,\beta,\gamma)) \vee (\alpha=\beta=\gamma \,\&\, A(\alpha,\beta,\gamma))], \end{aligned}$$

a decomposition of A into five independent predicates.

First we extend the auxiliary mapping $\mapsto$ in a rather obvious way by two new clauses

(xi) $\underline{\forall}\alpha_1\epsilon t_1\ldots\underline{\forall}\alpha_p\epsilon t_p\exists X_{n,m,p,q}A(\alpha_1,\ldots,\alpha_p,X) \mapsto$

$$\exists e\exists Y_{n+1,m,p,q}\forall n(en\neq 0\to \to \underline{\forall}\alpha_1\epsilon t_1*k_1^p n\ldots\underline{\forall}\alpha_p\epsilon t_p*k_p^p nA(\alpha_1,\ldots,\alpha_p,[1/en\dot{-}1]Y)),$$

where Y is a variable associated to X by a fixed correspondence, and

(xii) $\underline{\forall}\alpha_1\epsilon t_1\ldots\underline{\forall}\alpha_p\epsilon t_p\forall X_{n,m,p,q}A(\alpha_1,\ldots,X) \mapsto$

$$\forall X_{n,m,p,q}\underline{\forall}\alpha_1\epsilon t_1\ldots\underline{\forall}\alpha_p\epsilon t_pA(\alpha_1,\ldots,X).$$

The result of applying clauses (i) - (xii) as often as possible (i.e. till we have obtained a formula of degree 0) does not yet achieve a complete elimination of choice sequences - some formulae of the form

(1) $\underline{\forall}\alpha_1 \epsilon t_1 \ldots \underline{\forall}\alpha_p \epsilon t_p \; X_{n,m,q,r}(\mathfrak{A})$

may occur.

Let us consider a simple instance of (1), of the form $\underline{\forall}\alpha_1 \epsilon t_1 \underline{\forall}\alpha_2 \epsilon t_2 \; X_{0,0,0,2}(\alpha_1,\alpha_2)$. In a first crude approximation, we might be tempted to use a new variable $\overline{X}_{2,0,0,0}$ associated with $X_{0,0,0,2}$ by a fixed correspondence, to replace $\underline{\forall}\alpha_1 \epsilon t_1 \underline{\forall}\alpha_2 \epsilon t_2 X(\alpha_1,\alpha_2)$ by $\overline{X}(t_1,t_2)$, and to replace quantifiers $\forall X, \exists X$ by $\forall \overline{X}, \exists \overline{X}$. This procedure is too crude[†] however, since $\underline{\forall}\alpha_1 \epsilon n_1 \underline{\forall}\alpha_2 \epsilon n_2 X(\alpha_1,\alpha_2)$ does not correspond to an arbitrary predicate with two numerical arguments n_1,n_2; e.g. if we put $A(n_1,n_2) \equiv \underline{\forall}\alpha_1 \epsilon n_1 \underline{\forall}\alpha_2 \epsilon n_2 X(n_1,n_2)$, then A satisfies:

(i) $A(n_1,n_2) \rightarrow A(n_1 * m_1, n_2 * m_2)$ and

(ii) $\exists e \forall m (em \neq 0 \rightarrow A(n_1 * k_1^1 m, n_2 * k_2^2 m)) \rightarrow A(n_1,n_2)$,

which is certainly not generally the case for binary predicates of natural numbers.

Our problem is solved if we can find a *definable* operation Φ on arbitrary predicates $A(n_1,\ldots,n_p)$ such that Φ maps the predicates $A(n_1,\ldots,n_p)$ *onto* the predicates $B(n_1,\ldots,n_p)$ which are equivalent[††] to predicates of the form $\underline{\forall}\alpha_1 \epsilon n_1 \ldots \underline{\forall}\alpha_p \epsilon n_p C(\alpha_1,\ldots,\alpha_p)$. For then we can replace, returning to our example above, $\underline{\forall}\alpha_1 \epsilon t_1 \underline{\forall}\alpha_2 \epsilon t_2 X_{0,0,0,2}(\alpha_1,\alpha_2)$ by $(\Phi\overline{X})(t_1,t_2)$, and $\forall X, \exists X$ by $\forall \overline{X}, \exists \overline{X}$.

[†]The procedure described in Kreisel (1968), pages 239-240 is adequate for the first half of the elimination theorem (3.14) but not for extending the second half (3.15) to $\underset{\sim}{\mathrm{LSS}}$.

[††]Actually, our problem would already be solved on the weaker assumption that these predicates are a definable (in $\underset{\sim}{\mathrm{IDBS}}_1$) subset of all p-ary predicates of natural numbers.

3.17.

Now we proceed to construct such a Φ as required.

Definition. $A(n_1,\ldots,n_p)$ is said to be *progressive* with respect to $n_1,\ldots,n_p$ iff

(i) $A(n_1,\ldots,n_p) \to A(n_1 * m_1,\ldots,n_p * m_p)$,

(ii) $\exists e \forall m(em \neq 0 \to A(n_1 * k_1^p m,\ldots,n_p * k_p^p m)) \to A(n_1,\ldots,n_p)$.

Lemma A. If $A'(n_1,\ldots,n_p) \equiv \underline{\forall}\alpha_1 \in n_1 \ldots \underline{\forall}\alpha_p \in n_p A(\alpha_1,\ldots,\alpha_p)$, then A' is progressive with respect to $n_1,\ldots,n_p$. The proof is left as an exercise to the reader.

Note that

$$(\#(\alpha_1,\ldots,\alpha_p) \to A(\alpha_1,\ldots.\alpha_p)) \leftrightarrow \exists n_1 \ldots n_p(\#(\alpha_1,\ldots,\alpha_p) \to \alpha_1 \in n_1 \ \& \ldots \& \ \alpha_p \in n_p \ \& \ A'(n_1,\ldots,n_p)).$$

Lemma B. Suppose $B(n_1,\ldots,n_p)$ to be progressive with respect to $n_1,\ldots,n_p$. Then

$$\underline{\forall}\alpha_1 \in n_1 \ldots \underline{\forall}\alpha_p \in n_p[\exists n_1' \ldots n_p'(\alpha_1 \in n_1' \ \& \ldots \& \ \alpha_p \in n_p' \ \& \ \& \ B(n_1',\ldots,n_p'))] \leftrightarrow B(n_1,\ldots,n_p).$$

The proof of lemma B is left to the reader.

Lemma C. Let $A(n_1,\ldots,n_p)$ be any predicate, then we put

$$A^0(n_1,\ldots,n_p) \equiv_{\text{def}} \exists n_1' \ldots n_p'(n_1' \preceq n_1 \& \ \ldots \ \& n_p' \preceq n_p \ \& \ A(n_1',\ldots,n_p')),$$

$$A^*(n_1,\ldots,n_p) \equiv_{\text{def}} \exists e \forall m(em \neq 0 \to A^0(n_1 * k_1^p m,\ldots,n_p * k_p^p m)).$$

Then $A^*(n_1,\ldots,n_p)$ is progressive with respect to $n_1,\ldots,n_p$.

Proof: We have to show that clauses (i) and (ii) in the definition of progressiveness are satisfied.

Assume $A^*(n_1,\ldots,n_p)$; this implies for some f

$$\forall m(fm \neq 0 \to A^0(n_1 * k_1^p m,\ldots,n_p * k_p^p m)).$$

From the definition of $f[m_1,\ldots,m_p]$

$$\forall m((f[m_1,\ldots,m_p])m \neq 0 \to A^0(n_1 * m_1 * k_1^p m,\ldots,n_p * m_p * k_p^p m)),$$

where we have made use of the fact that

$$A^0(n_1',\ldots,n_p') \to A^0(n_1' * m_1',\ldots,n_p' * m_p'),$$

hence especially

$$A^0(n_1 * k_1^p m',\ldots,n_p * k_p^p m') \to A^0(n_1 * m_1 * k_1^p m,\ldots,n_p * m_p * k_p^p m)$$

for an m' such that $k_i^p m' \preceq m_i * k_i^p m (1 \leq i \leq p)$.

Next let us assume for some f

$$\forall m(fm \neq 0 \to \exists e \forall n(en \neq 0 \to A^0(n_1 * k_1^p m * k_1^p n,\ldots,n_p * k_p^p m * k_p^p n))).$$

For any e, let us define e^m by

$$\begin{cases} e^m n = 0 \text{ for } n \prec m \\ e^m(m*n) = en \\ e^m n = 1 \text{ in all other cases.} \end{cases}$$

Since $e^m \in K$, as is easily seen, we conclude that

$$\forall m(fm\neq 0 \to \exists e \forall n(e(m*n)\neq 0 \to A^0(n_1*k_1^p m*k_1^p n,\ldots))).$$

Now we apply AC-NF and find an e' such that

$$\forall m(fm\neq 0 \;\&\; e'(\hat{m}*m*n)\neq 0 \to A^0(n_1*k_1^p(m*n),\ldots,n_p*k_p^p(m*n))).$$

Now take $f' = f/e'$ (cf. Kreisel and Troelstra (1970), 3.2.3-3.2.5, 3.2.8-3.2.9), then if $f'n\neq 0$, also $fn\neq 0$, $e'(\langle h(f,n)\dot{-}1\rangle *n) \neq 0$, and therefore (since $fn\neq 0 \to h(f,n)\neq 0$ & $h(f,n)\dot{-}1 \preceq n$ & $f(h(f,n)\dot{-}1)=fn$ it follows that

$$f(h(f,n)\dot{-}1)\neq 0 \;\&\; e'(\langle h(f,n)\dot{-}1\rangle *(h(f,n)\dot{-}1)*n')\neq 0 \to$$
$$\to A^0(n_1*k_1^p n,\ldots,n_p*k_p^p n)$$

where n' is such that $(h(f,n)\dot{-}1)*n' = n$. Therefore

$$\forall n(f'n\neq 0 \to A^0(n_1*k_1^p n,\ldots,n_p*k_p^p n)),$$

and thus $A^*(n_1,\ldots,n_p)$.

Summarizing, we have shown that the predicates B: $B(n_1,\ldots,n_p) \equiv_{\text{def}} \underline{\forall}\alpha_1\in n_1\ldots\underline{\forall}\alpha_p\in n_p A(\alpha_1,\ldots,\alpha_p)$ are exactly the predicates progressive in $n_1,\ldots,n_p$ and the definable operation Φ which assigns A^* to A in Lemma C maps arbitrary predicates in $n_1,\ldots,n_p$ onto the predicates progressive in $n_1,\ldots,n_p$.

Now we can show how to complete the translation. Consider a sub-formula of the form (taking a typical case)

(1) $\underline{\forall}\alpha_1\in t_1\underline{\forall}\alpha_2\in t_2\; X_{1,1,1,2}(t[\alpha_1,\alpha_2],\phi_1[\alpha_1,\alpha_2],\phi_2[\alpha_1,\alpha_2],\alpha_1,\alpha_2)$.

In the functors ϕ_1, ϕ_2 the variables α_1,α_2 occur only in sub-terms, i.e. there are $\phi_1'[x_1,\ldots,x_n]$, $\phi_2'[y_1,\ldots,y_m]$ (ϕ_1', ϕ_2' not containing lawless variables) and terms $s_1[\alpha_1,\alpha_2],\ldots,s_n[\alpha_1,\alpha_2],r_1[\alpha_1,\alpha_2],\ldots,r_m[\alpha_1,\alpha_2]$ such that $\phi_1[\alpha_1,\alpha_2] \equiv \phi_1'[s_1,\ldots,s_n]$, $\phi_2[\alpha_1,\alpha_2] \equiv \phi'[r_1,\ldots,r_m]$. (We may suppose $s_1,\ldots,s_n$, $r_1,\ldots,r_m$ to be chosen so as to have minimal complexity and to be non-overlapping.) We now replace (1) by

(2) $\forall n_1 n_2 \forall x \forall y_1 \ldots y_n \forall z_1 \ldots z_m [\{\forall a_1 \epsilon t_1 * n_1 \forall a_2 \epsilon t_2 * n_2 (t[a_1, a_2] = x)$ &
$\quad$ & $\forall a_1 \epsilon t_1 * n_1 \forall a_2 \epsilon t_2 * n_2 (s_1[a_1, a_2] = y_1$ &...& $s_n[a_1, a_2] = y_n)$ &
$\quad$ & $\forall a_1 \epsilon t_1 * n_1 \forall a_2 \epsilon t_2 * n_2 (r_1[a_1, a_2] = z_1$ &...& $r_m[a_1, a_2] = z_m)\} \rightarrow$
$\rightarrow \underline{\forall} \alpha_1 \epsilon t_1 * n_1 \underline{\forall} \alpha_2 \epsilon t_2 * n_2 X_{1,1,1,2}(x, \phi'_1[y_1, \ldots, y_n], \phi'_2[z_1, \ldots, z_m], \alpha_1, \alpha_2)]$
and then (abbreviating (3.2) as
$\forall n_1 n_2 \forall x \forall y_1 \ldots y_n \forall z_1, \ldots, z_m [F_1 \rightarrow F_2]$) we replace F_2 in (3.2) by

$$\overline{X}^*_{3,1,1,0}(n_1, n_2, x, \phi'_1[y_1, \ldots, y_n], \phi'_2[z_1, \ldots, z_m]),$$

where $\overline{X}$ is assigned to X by a fixed mapping, and $\overline{X}^*$ indicates the progressive closure with respect to the first and second numerical variables. This yields

(3) $\quad \forall n_1 n_2 \forall x \forall y_1 \ldots y_n \forall z_1 \ldots z_m [F_1 \rightarrow \overline{X}^*_{3,1,1,0}(n_1, n_2, x, \phi'_1, \phi'_2)]$;

and finally we replace every occurrence of $\forall X_{1,1,1,2}, \exists X_{1,1,1,2}$ by $\forall \overline{X}_{3,1,1,0}, \exists \overline{X}_{3,1,1,0}$ respectively.

3.18.

In order to extend the final part of the elimination theorem to $\underset{\sim}{\text{LSS}}$*, we must correspondingly extend $\underset{\sim}{\text{IDB}}_1$ to a system $\underset{\sim}{\text{IDBS}}_1$, containing variables $X_{n,m,p}$ for species with n numerical, m constructive function, and p K-function arguments, full comprehension for such species, and an axiom of choice

AC-NS $\quad \forall x \exists X_{n,m,p} A(x, X) \rightarrow \exists Y_{n+1,m,p} \forall x A(x, [1/x]Y)$.

3.19 - 3.23. Reduction of $\underset{\sim}{\text{LS}}$ to the theory of a single lawless sequence.

3.19.

Intuitively, one thinks of the universe of lawless sequences as being non-enumerable. And in fact, it is easy to show that the assumption that an enumerating predicate A exists such that

(1) $\quad \forall \alpha \exists x \forall y z (\alpha y = z \leftrightarrow A(x, y, z))$

conflicts with LS3. On the other hand, it is possible to construct a model for $\underset{\sim}{\text{LS}}$ from a single lawless sequence; this model is countable. The idea of the model may be described as follows.

We note that there is only one existential axiom (LS1) for lawless sequences which can be satisfied by a countable

species of lawless sequences.
Next we recall the thought-experiment related in section 2.8, involving sequences α,β,γ such that $\gamma(2n) = \alpha n$, $\gamma(2n+1) = \beta n$; we had to choose whether to regard α,β as lawless (but then not γ) or we could regard γ as lawless (but not α,β). This suggests that we should consider the following species of sequences

$$\mathcal{U}_\alpha = \{n*(\alpha)_n : n \in N\}$$

where α is a 'fixed' lawless sequence, $(\alpha)_x = \lambda y.\alpha(x,y)$, and $n*(\alpha)_n$ is the 'concatenation' of n and $(\alpha)_n$. The $(\alpha)_x$, for distinct x, behave as being completely independent (i.e. they behave as independent lawless sequences) but of course we cannot refer to $(\alpha)_n$ and α as both being lawless within the same context. Similarly, if we consider sequences $n*(\alpha)_n$ instead of $(\alpha)_n$, they also behave as being independent, but in addition we have now made sure that each possible finite initial segment occurs among them, as required by LS1. As a result, we can show in $\underset{\sim}{\text{LS}}$ that if we let α range over $\mathcal{U}_\alpha$, and we interpret = between lawless sequences as equality between natural numbers, i.e. $n*(\alpha)_n = m*(\alpha)_m \equiv_{\text{def}} n = m$, then we obtain a model for $\underset{\sim}{\text{LS}}$.

3.20.

We shall indicate the idea of the proof. To state the assertion from which our theorem will follow, we introduce the projections Π_n: for any finite sequence $\bar{a}x$, $\Pi_n(\bar{a}x)$ is obtained by concatenating n and the sequences of values of $(a)_n$ for arguments y such that $j(n,y) < x$ (i.e. the sequence of values of $(a)_n$ determined by $\bar{a}x$). We have assumed j to satisfy $x \le x'$ & $y \le y' \to j(x,y) \le j(x',y')$. We also define α^n:

$$\begin{cases} x < \text{lth}(n) \to \alpha^n(x) = (n)_x \\ x \ge \text{lth}(n) \to \alpha^n(x) = \alpha(n, x \dot{-} \text{lth}(n)) \end{cases}$$

and we define

$$\#(n_1,\ldots,n_p) \equiv_{\text{def}} \bigwedge_{1 \le i < j \le p} n_i \ne n_j \, .$$

We now introduce a syntactical mapping Γ.

Definition. Let $v_1, v_2, v_3, \ldots$ be a list of the numerical variables of $\underset{\sim\sim}{LS}$; let $w_1, w_2, w_3, \ldots$ be a list of the lawless variables of $\underset{\sim\sim}{LS}$.

Let Fm* denote the class of formulae of $\underset{\sim\sim}{LS}$ containing odd-numbered numerical variables only. We now define, with α a fixed lawless variable, a syntactical mapping Γ of Fm* into the formulae of (a conservative extension of) $\underset{\sim\sim}{LS}$ as follows:

(i) If $t_1 = t_2 \in \mathrm{Fm}^*$, then $\Gamma(t_1 = t_2)$ is obtained by replacing every occurrence of w_i by $\alpha^{v_{2i}}$, for all i.

(ii) Γ is a homomorphism for &, $\vee$, $\rightarrow$, ($\neg$), and quantifiers $\forall x$, $\exists x$, $\forall a$, $\exists a$, $\forall e$, $\exists e$ (i.e. $\Gamma(A \,\&\, B) \equiv \Gamma(A) \,\&\, \Gamma(B)$, etc).

(iii) $\Gamma(\forall w_i A) \equiv \forall v_{2i} \Gamma(A)$, $\Gamma(\exists w_i A) \equiv \exists v_{2i} \Gamma(A)$.

We shall call $\Gamma(A)$ the Γ-transform of A, and if $B(\beta_1, \ldots, \beta_p) \in \mathrm{Fm}^*$ we write $B^*(\underset{\sim}{n}_1, \ldots, \underset{\sim}{n}_p)$ for the Γ-transform of B (where $\underset{\sim}{n}_i = v_{2j}$ in case $\beta_i \equiv w_j$).

Now we can prove the following theorem by induction on the complexity of B:

3.21.

Theorem. Let B be any formula of Fm*. Then in $\underset{\sim\sim}{LS}$

$$\forall u \forall n_1 \ldots n_p (\#(n_1, \ldots, n_p) \rightarrow \{\forall \alpha \in u\; B^*_\alpha(\underset{\sim}{n}_1, \ldots, \underset{\sim}{n}_p) \leftrightarrow$$
$$\leftrightarrow \underline{\forall} \beta_0 \in \Pi_{n_1} u \ldots \underline{\forall} \beta_p \in \Pi_{n_p} u B(\beta_1, \ldots, \beta_p)\}) \quad (u \text{ not free in } B, B^*_\alpha).$$

Proof: See Troelstra (1970A), 2.3.

Corollary. A sentence of Fm* is valid if its Γ-transform is valid for some α; if F is valid, then $\forall \alpha \Gamma(F)$ is valid.

3.22.

Theorem. If $\underset{\sim\sim}{LS}^1$ is the fragment of $\underset{\sim\sim}{LS}$ dealing with formulae containing at most one free lawless variable (axiomatized by dropping LS2 and restricting LS3 and LS4 to the one-variable case), then

$$\underset{\sim\sim}{LS} \vdash B \Leftrightarrow \underset{\sim\sim}{LS}^1 \vdash \forall \alpha B^*_\alpha$$

(B not containing lawless variables free).

Proof: Let $\rho(B) \equiv \forall\alpha B^*_\alpha$. Obviously, $\underset{\sim}{LS}^1 \vdash \rho(B)$ $\Leftrightarrow \underset{\sim}{IDB}_1 \vdash \tau\rho(B)$; in combination with $\underset{\sim}{IDB}_1 \vdash \tau(B) \leftrightarrow \tau\rho(B)$ (3.3, 3.21) and $\underset{\sim}{LS} \vdash A \Leftrightarrow \underset{\sim}{IDB}_1 \vdash \tau(B)$ (3.3), the theorem follows.

3.23.

The theory of a single lawless sequence (or the theory of predicates with a single lawless parameter) is of interest in connection with completeness for intuitionistic predicate logic (see Chapter 7).

The theory of lawless sequences ranging (cf. section 2.18) over a finitely branching tree ('fan') may be developed parallel to the theory of lawless sequences ranging over Baire space. However, it is possible to construct a model for such lawless sequences from the lawless sequences ranging over Baire space as follows. Let a be a bounding function (cf. section 2.18) for the sequences considered. Let Γ_a be a mapping of type $N^N \to N^N$, specified by a neighbourhood function ϕ such that

$$\begin{cases} \phi(\langle\rangle) = 0 \\ \phi(\hat{x} * \bar{\alpha}y) = 0 \text{ if } y \le x \\ \phi(\hat{x} * \bar{\alpha}y) = \min[\alpha x, ax] \text{ if } x < y. \end{cases}$$

Then $\{\phi | \alpha : \alpha \in LS\}$ is a model for the theory of lawless sequences bounded by a. (Cf. section B7 and the references given there).

4
PROJECTIONS OF LAWLESS SEQUENCES

4.1. *Introduction.*

For the development of mathematical theories logically based on analysis, i.e. on natural numbers and sequences of natural numbers, the domain of sequences of natural numbers should have certain properties not possessed by lawless sequences, notably closure under certain non-trivial continuous operations. One of the simplest illustrations is provided by the theory of Baire space (classically defined as N^N provided with the product topology if N is provided with the standard discrete topology). If the elements of Baire space range over lawless sequences, the structure of the family of open sets is the same as in the classical case, but there are no autohomeomorphisms except the identity. So if we wish to have a space more closely resembling Baire space the sequences of N^N should range over a domain closed under (some) continuous operations.

We are led to similar requirements in the case of the more familiar example of real-valued analysis. If we wish to develop real-valued analysis based on a certain domain $\mathscr{A}$ of sequences of natural numbers, this can be done in various ways; one of the most obvious methods is as follows: let $\langle r_n \rangle_n$ be a standard (lawlike) enumeration of the rationals; the real-number generators relative to $\mathscr{A}$ (or $\mathscr{A}$-r.n.g.'s) are sequences $\langle r_{\alpha n} \rangle_n, \alpha \in \mathscr{A}$ such that $\forall k \exists m \forall n_1 n_2 (|r_{\alpha(m+n_1)} - r_{\alpha(m+n_2)}| < 2^{-k})$, and equivalence ($\sim$) is defined, as usual, by

$$\langle r_{\alpha n} \rangle_n \sim \langle r_{\beta n} \rangle_n \equiv_{\text{def}} \forall k \exists n \forall m (|r_{\alpha(n+m)} - r_{\beta(n+m)}| < 2^{-k}),$$

Let x_α denote the real number corresponding to $\langle r_{\alpha n} \rangle_n$. Now obviously, to get a suitable class of $\mathscr{A}$-reals, $\mathscr{A}$ should satisfy certain conditions; e.g. the requirement that $\langle r_{\alpha n} \rangle_n$ is a r.n.g. imposes certain restrictions on α. Similarly, the condition that the $\mathscr{A}$-reals are closed under addition requires

$\forall\alpha\forall\beta\exists\gamma(x_\alpha+x_\beta = x_\gamma)$ which is fulfilled for example if

$$\forall\alpha\forall\beta\exists\gamma(\langle r_{\alpha n}+r_{\beta n}\rangle_n = \langle r_{\gamma n}\rangle_n).$$

Of course, some of the requirements may be eliminated by choosing a different way of associating r.n.g.'s to elements of $\mathscr{A}$; for example, if we use a lawlike function a such that $\lambda m.a(n,m,k))$ enumerates the set $\{m:|r_n-r_m|<2^{-k}\}$, we can associate to each α a r.n.g. $\langle r_{\phi n}\rangle_n$ by requiring $\forall n(\phi(n+1) = a(\phi n, a(n+1), n+1))$.But then e.g. closure under addition still imposes certain restrictions.

In fact, closure under some non-trivial continuous operations *has* to be satisfied by $\mathscr{A}$, if we wish to base real-valued analysis on $\mathscr{A}$; this is illustrated by the following proposition (proof of which is left to the reader): Let a fixed lawlike operation assign a r.n.g. $\langle r_n^\alpha\rangle_n$ to each lawless α; let x_α be the corresponding real. Then condition $\forall\alpha\exists\beta(2x_\alpha = x_\beta)$ implies that the set of the x_α is enumerable and consists of lawlike elements only.

One of the most natural conditions to impose on $\mathscr{A}$ is closure under all (lawlike) continuous operations, although this is perhaps more than is necessary. However, closure under continuous operations Γ which map reals into reals, i.e. satisfying $x_\alpha = x_\beta \rightarrow x_{\Gamma\alpha} = x_{\Gamma\beta}$ is not enough, as can be seen from the following example. There does exist a continuous operation Γ such that

(1) $\quad \forall x_\alpha(x_{\Gamma\alpha}^3 - 3x_{\Gamma\alpha} + x_\alpha = 0)$

(in words: the equation $y^3 - 3y + c = 0$ has a root x for all values of the parameter c, which can be approximated from an approximation of c), but there is no Γ satisfying (1) and in addition $x_\alpha = x_\beta \rightarrow x_{\Gamma\alpha} = x_{\Gamma\beta}$.

For such a Γ, if existing, should be continuous in the usual topology for the reals; and that such a Γ does not exist is easily seen from the behaviour of the equation $x^3 - 3x + c = 0$ for varying c: for $c < -2$ there is only one root > 1; for $c = -2$ there is a double root -1, and a root >1; for $-2 < c < 2$ there are three roots, x_0, x_1, x_2 satisfying $x_0 < -1$, $-1 < x_1 < 1, 1 < x_2$; for $c = 2$ there is one root < -1, and a double root 1; for $c > 2$ there is only one root < -1. If we try to pick a root for each c, we cannot do so continuously, we somewhere have to 'jump'.

We also wish $\mathscr{A}$ to satisfy some continuity axioms, since the continuity properties of choice sequences give mathematics based on choice sequences its distinctive flavour (cf. our remarks in section 2.15); continuity properties which themselves are incompatible with classical logic compensate in some cases for the absence of the excluded third (cf. the examples taken from mathematics in Chapter 6).

The preceding discussion therefore induces us to look for universes $\mathscr{A}$ satisfying (i) closure under all lawlike continuous operations and (ii) some interesting continuity axioms. Since the lawless sequences themselves obviously do not satisfy (i), although they do satisfy (ii), we may try to construct universes $\mathscr{A}$ from the lawless sequences. In this section we shall discuss two examples of such universes constructed from lawless sequences.

The first example, $\mathscr{U}$, is discussed in sections 4.2 - 4.15, and consists of all sequences obtained by applying all lawlike continuous operations to all u-tuples of lawless sequences, for all u. The principal properties of $\mathscr{U}$ are $\forall\alpha\exists!x$-continuity and $\forall x\exists y$-choice; on the other hand, $\forall\alpha\exists x$-continuity and $\forall x\exists\alpha$-choice are not generally valid. The other example consists of a whole collection of universes $\mathscr{U}^*_\alpha$ (α a lawless sequence); $\mathscr{U}^*_\alpha$ is obtained by applying all lawlike continuous operations to a single lawless sequence α. Bar induction holds for all $\mathscr{U}^*_\alpha$, and so does $\forall x\exists\alpha$-choice; but continuity has been eatablished only under special conditions. Validity in all $\mathscr{U}^*_\alpha$ is equivalent to validity in the topological model for intuitionistic analysis described in Moschovakis (1973), provided the elements of Baire space are interpreted as ranging over lawless sequences. These examples are not only interesting as a basis for intuitionistic versions of well-known mathematical theories such as real-valued analysis, but are also of interest in other respects, for example as a means of obtaining consistency and formal independence results (cf. e.g. 4.10 - 4.14).

4.2. - 4.15. The universe $\mathcal{U}$.

4.2.

The first universe of sequences constructed from lawless sequences which we shall study is defined as $(e|(\varepsilon_1,\ldots,\varepsilon_u)$ abbreviating $e|\nu_u(\varepsilon_1,\ldots,\varepsilon_u))$

$$\mathcal{U} = \{\alpha : \exists e \underline{\exists}\varepsilon_1 \ldots \underline{\exists}\varepsilon_u(\alpha = e|(\varepsilon_1,\ldots,\varepsilon_u), u \in N - \{0\})\}$$

$(\varepsilon_1,\ldots,\varepsilon_u$ variables for lawless sequences. In discussing $\mathcal{U}$, we shall reserve α,β,γ as variables for elements of $\mathcal{U}$).

The following schemata can be shown to hold for $\mathcal{U}$ (with respect to extensional predicates)

AC-NN* $\forall x \exists y A(x,y,\alpha) \to \exists \beta \forall x A(x,\beta x,\alpha)$,

WC-N!* $\forall \alpha \exists ! x A(\alpha,x,\beta) \to \forall \alpha \exists x \exists y \forall \gamma \in \overline{\alpha} y A(\gamma,x,\beta)$

(in fact, in sections 4.7 and 4.9 much more is established) and also

C-N!* $\forall \alpha \exists ! x A(\alpha,x,\beta) \to \exists \gamma \in K_0 \forall \alpha A(\alpha,\gamma(\alpha),\beta)$,

where K_0 is defined as $K_0\gamma \equiv \forall \alpha \exists x(\gamma(\overline{\alpha}x) \neq 0)$ & $\forall nm(\gamma n \neq 0 \to \gamma n = \gamma(n * m))$.

SBC $\forall \alpha \exists x A(\overline{\alpha}x) \to \exists e \forall \alpha A(\overline{\alpha}(e(\alpha)))$

(where e,f as before are used for variables ranging over K-functions).

As a consequence of SBC and WC-N!* we have

BC-N! $\forall \alpha \exists ! x A(\alpha,x) \to \exists e \forall \alpha A(\alpha,e(\alpha))$

and, as a consequence of a generalization of SBC with parameters, the schema of bar induction

BI* $[\forall \alpha \exists x P(\overline{\alpha}x,\beta)$ & $\forall n(P(n,\beta) \to Q(n,\beta))$ & $\forall nm(P(n,\beta) \to$
$\to P(n*m,\beta))$ & $\forall n(\forall y Q(n*\hat{y},\beta) \to Q(n,\beta))] \to Q(0,\beta)$.

As will be seen from the proofs, intuitively these schemata are justified for any extensional A,P,Q; the proofs can actually be carried out in a system $\underset{\sim}{\mathrm{CL}}$ ('constructs of lawless sequences'), a suitable extension of $\underset{\sim}{\mathrm{LS}}$.

<u>Remark on the designations of the schemata.</u> Our usage here conforms as far as possible to Kreisel and Troelstra (1970), with one exception noted below. 'AC', 'WC', 'C', 'SBC', 'BC', 'BI' stand for 'Axiom of Choice', 'Weak Continuity', 'Continuity', 'Special Bar Continuity', 'Bar Continuity', and 'Bar Induction' respectively. Later on we shall encounter 'RDC' and 'A' for 'Relativized Dependent Choices'

and 'Analyticity' respectively. In the combinations of letters behind the hyphen (when present) 'N', 'F', 'C', 'S' stand for 'Numbers', 'lawlike Functions', 'Choice sequences' and 'Species' respectively.

! refers to an existential quantifier with uniqueness condition, and for schemata involving choice sequences * indicates the schema *with* choice parameters; for the schema without choice parameters the * is dropped. So for example AC-NC* indicates the schema

AC-NC* $\forall x \exists \alpha A(x,\alpha,\beta) \rightarrow \exists \gamma \forall x A(x,(\gamma)_x,\beta)$.

(In Kreisel and Troelstra (1970) this schema was denoted by AC-NC; the reason was that the schema

$$\forall x \exists \alpha A(x,\alpha) \rightarrow \exists \gamma \forall x A(x,(\gamma)_x)$$

(without parameters) was redundant since it was contained in

AC-NF $\forall x \exists a A(x,a) \rightarrow \exists c \forall x A(x,(c)_x)$

in the presence of the 'specialization principle' $\exists \alpha A\alpha \rightarrow \exists a A a$. However, we regard our present use of * as more consistent.)

Another point worth mentioning is the following. Relative to a system of analysis (such as EL) where the function variables may be interpreted as either ranging over lawlike or over choice sequences, it is rather indifferent whether we write AC-NF or AC-NC*. But we shall, for consistency, adhere to the practice that where 'lawlike' variables (a,b,c,d) are used we write AC-NF, and where 'choice' variables $(\alpha,\beta,\gamma,\delta)$ are used we write AC-NC*; and similarly for other schemata. An *important* convention is that (as before) choice parameters are always supposed to be *shown*. Now we turn to the description of CL.

4.3. *Description of* CL.

We use ε,η,ζ as variables for u-tuples of lawless sequences; α,β,γ as variables for elements of $\mathcal{U}$; numerical variables, variables for lawlike sequences, and elements of K as in IDB_1. Rules of term formation are similar to those in LS, where ε,η,ζ are treated as lawless variables,

α,β,γ as lawlike variables. Constants etc. as in $\underset{\sim}{\mathrm{IDB}}_1$. Moreover there is an operation $||$, the rank- or 'tuplicity' operator: $|\varepsilon|=u$ means that ε is a (coding of) *a* $(u+1)$-tuple of independent lawless sequences. With respect to natural numbers, lawlike functions, and K-functions we assume the axioms and schemata of $\underset{\sim}{\mathrm{IDB}}_1$; also with respect to the other sorts of variables we adopt the usual quantifier rules and axioms.

Let us introduce as abbreviations

$$\forall\varepsilon^n A(\varepsilon^n) \equiv_{\mathrm{def}} \forall\varepsilon(|\varepsilon|=n \rightarrow A(\varepsilon))$$
$$\exists\varepsilon^n A(\varepsilon^n) \equiv_{\mathrm{def}} \exists\varepsilon(|\varepsilon|=n \;\&\; A(\varepsilon)).$$

Moreover, let ν_u be the usual coding of u-tuples of sequences, viewed as a function of u also, and j^u_k its inverses (also viewed as functions depending on u,k) such that $j^u_k = j^u_u$ for $k \geq u$. We shall assume $\nu_{u+1}(\chi_0,\chi_1,\ldots,\chi_u) = j(\chi_0,\nu_u(\chi_1,\ldots,\chi_u))$, so $j^u_1\chi = j_1\chi$.

<u>Definition</u>. $D(\varepsilon,\eta) \equiv_{\mathrm{def}} \forall n\leq|\varepsilon|\forall m\leq|\eta|(j^{|\varepsilon|+1}_{n+1}\varepsilon \neq j^{|\eta|+1}_{m+1}\eta)$ (i.e. the lawless sequences contained in ε,η are independent). Now we add two axioms to express the 'tuple-character' of the range of ε,η etc.:

CL1 $\quad |\varepsilon|=k \rightarrow \forall n\leq k\exists\eta^0(\eta^0 = j^{k+1}_{n+1}\varepsilon)\&\forall nm(n<m\leq k-1 \rightarrow j^{k+1}_{n+1}\varepsilon \neq j^{k+1}_{m+1}\varepsilon)$

CL2 $\quad D(\varepsilon^0,\eta^x) \rightarrow \exists\zeta^{x+1}(j_1\zeta^{x+1} = \varepsilon^0 \;\&\; j_2\zeta^{x+1} = \eta^x).$

CL1 expresses the fact that all ε are u-tuples for some u, and and CL2 expresses the fact that all u-tuples of lawless sequences are in the range ε,η, etc. The next four axioms are a straightforward exposition of the axiom schemata LS1-4 of $\underset{\sim}{\mathrm{LS}}$:

CL3 $\quad \forall n\exists\varepsilon^0(\varepsilon^0 \in n)$

CL4 $\quad \varepsilon^0=\eta^0 \vee \varepsilon^0\neq\eta^0$

CL5 $\quad A\varepsilon^n \rightarrow \exists m(\varepsilon^n \in m \;\&\; \forall\eta^n \in m\; A(\eta^n))$

CL6 $\quad \forall\varepsilon^n\exists a A(\varepsilon^n,a) \rightarrow \exists e\exists b\forall\varepsilon^n A(\varepsilon^n,(b)_{e(\varepsilon^n)}).$

(Here, as before in $\underset{\sim}{\mathrm{LS}}$, we have adopted the convention that all variables for u-tuples and elements of $\mathcal{U}$ occurring free are actually shown.)

Finally, we need an axiom to determine the connection between lawless sequences and $\mathcal{U}$:

CL7 $\forall\alpha\exists e\exists\varepsilon(\alpha=e|\varepsilon)\ \&\ \forall\varepsilon\forall e\exists\alpha(\alpha=e|\varepsilon)$.

Some consequences of CL1-7 are (proofs either obvious or very similar to proofs in LS)

(1) $\forall n\exists\varepsilon^m(\varepsilon^m\in n)$

(2) $D(\varepsilon,\eta)\ \vee\ \ D(\varepsilon,\eta)$

(3) $\forall m'(\forall\varepsilon^n\in m*m'A\varepsilon^n\to\forall\varepsilon^n\in m*m'B\varepsilon^n)\leftrightarrow\forall\varepsilon^n\in m(A\varepsilon^n\to B\varepsilon^n)$

(4) $\forall\varepsilon^n\in m\exists xA(\varepsilon^n,x)\to\exists e\forall\varepsilon^n\in mA(\varepsilon^n,e(\varepsilon^n))$.

Remark. The expressive power of CL is greater than that of LS since we now have quantifiers over finite sequences (of variable length) of lawless sequences. On the other hand, the intuitive justification of LS provides us at the same time with an intuitive justification for CL.

In establishing the validity of various schemata for $\mathcal{U}$ we shall not formalize the proofs in CL (since such a complete formalization in CL starting from the proofs given below is straightforward). The same proofs but formalized to a much greater degree may be found in Troelstra(1969A), in which the results on $\mathcal{U}$ were first proved. Actually, the results as formulated here slightly extend the results in Troelstra (1969A): the schemata WC-N! and BI are here extended to the corresponding schemata WC-N!*, BI* containing additional parameters for elements of $\mathcal{U}$. Especially with respect to the validity of WC-N!* it is useful to give the proof here in a much more informal style with a correspon-ding gain in perspicuity. It should be noted that we in fact establish continuity of mappings from complete separ-able metric spaces into separable metric spaces; WC-N!* is only a special case.

4.4.

Theorem. AC-NN* holds for $\mathcal{U}$. (All instances of AC-NN* are provable in CL.)

Proof: Let $\forall x\exists yA(x,y,\alpha)$. Then for some e, and p-tuple ε^p, $\alpha=e|\varepsilon^p$. It follows that for some n (CL5) $\varepsilon^p\in n\ \&\ \forall\eta^p\in n\forall x\exists yA(x,y,e|\eta^p)$, hence

$$\forall x\forall\eta^p\in n\exists yA(x,y(e|\eta^p)$$

and therefore ((4) in 4.3)

$$\forall x \exists f \forall \eta^P \in n A(x, f(\eta^P), e|\eta^P)$$

and by AC-NF for lawlike sequences, for some f'

$$\forall x \forall \eta^P \in n\ A(x, \lambda n. f'(\hat{x} * n)(\eta^P), e|\eta^P).$$

Then

$$\forall x A(x, \lambda n. f'(\hat{x} * n)(\varepsilon^P), e|\varepsilon^P)$$

and if we put $\beta = f'|\varepsilon^P$,

$$\forall x A(x, \beta x, \alpha).$$

4.5.

Our next aim is to establish WC-N!*. We shall in fact prove a more general theorem, for which purpose we must introduce some definitions first.

Definition. A *metric space* $\langle V, \rho\rangle$ consists of a species V and a mapping $\rho : V \times V \to \underset{\sim}{R}$ ($\underset{\sim}{R}$ species of reals, which of course is also defined relative to some universe of number theoretic functions) such that

(1) $\rho(x,y) \nless 0$, $\rho(x,x) = 0$

(2) $\rho(x,y) = \rho(y,x)$

(3) $\rho(x,y) \ngtr \rho(x,z) + \rho(z,y)$.

A *separable* metric space is a metric space $\langle V, \rho\rangle$ with a sequence of *basis* points $\langle p_n\rangle_n$ such that $\rho(p_n, p_m) = 0 \vee \rho(p_n, p_m) \# 0$ for all n, m, $\langle p_n\rangle_n \subset V$, and which is dense in the space, i.e. for each $x \in V$ there is a sequence $\langle p_{\alpha n}\rangle_n$ such that $\lim_n p_{\alpha n} = x$ (i.e. $\lim_n \rho(p_{\alpha n}, x) = 0$). $\langle p_n\rangle_n$ is said to be a *basis* for the space.

A *complete* metric space $\langle V, \rho\rangle$ is a metric space such that for each sequence $\langle x_n\rangle_n \subset V$ such that $\forall k \exists n \forall m_1 m_2 (m_1 \geq n \ \&\ m_2 \geq n \to \rho(x_{m_1}, x_{m_2}) < 2^{-k})$ there is a $y \in V$ such that $\lim_n (x_n) = y$.

Additional conventions for $\mathcal{U}$. In our discussion of $\mathcal{U}$ we shall assume our separable metric spaces $\langle V, \rho\rangle$ with basis $\langle p_n\rangle_n$ to be specified by a *lawlike* metric, i.e. there is a lawlike function a, such that relative to a fixed lawlike enumeration of the rationals (e.g. the standard one) $\langle r_n\rangle_n$

$$|r_{a(n,m,k)} - \rho(p_n, p_m)| < 2^{-k}.$$

Moreover, in a separable metric space (defined relative to the universe $\mathcal{U}$) a sequence $\langle x_n \rangle_n$ is supposed to be specified by a sequence $\alpha \in \mathcal{U}$ such that $x_n = \lim_m p_{\alpha(n,m)}$.

4.6.

Definition. Let $\chi_1,\ldots,\chi_k$ be used for arbitrary sequences. We define

$$e[\chi_1,\ldots,\chi_p]n=y+1 \equiv_{def}$$
$$\equiv \exists m(\chi_1 \in k_1^{p+1}m \ \&\ldots\&\ \chi_p \in k_p^{p+1}m \ \&\ n=k_{p+1}^{p+1}m \ \&\ em=y+1)$$

and

$$e[\chi_1,\ldots,\chi_p](\chi)=y \equiv_{def} \exists z(e[\chi_1,\ldots,\chi_p](\bar{\chi}z)=y+1).$$

Note that

$$\forall\chi\exists z(e[\chi_1,\ldots,\chi_p](\bar{\chi}z)\neq 0)$$

(cf. proof in section 3.9), and also

$$\forall nm(e[\chi_1,\ldots,\chi_p]n\neq 0 \rightarrow e[\chi_1,\ldots,\chi_p](n)=e[\chi_1,\ldots,\chi_p](n*m)).$$

Lemma. Let $\varepsilon^0, \varepsilon^p, \eta^p$ be variables as in CL. Then

$$\forall\varepsilon^0(D(\varepsilon^0,\varepsilon^p) \rightarrow \exists x A(\varepsilon^0,\varepsilon^p\ x)) \rightarrow$$
$$\rightarrow \exists e\forall\varepsilon^0(D(\varepsilon^0,\varepsilon^p) \rightarrow A(\varepsilon^0,\varepsilon^p,e[\varepsilon^p](\varepsilon^0))).$$

The corresponding formulation for LS is

$$\underline{\forall}\varepsilon\exists x A(\varepsilon,\varepsilon_1,\ldots,\varepsilon_p,x) \rightarrow \exists e\underline{\forall}\varepsilon A(\varepsilon,\varepsilon_1,\ldots,\varepsilon_p,e[\varepsilon_1,\ldots,\varepsilon_p](\varepsilon))$$

(where $\varepsilon,\varepsilon_1,\ldots,\varepsilon_p$ are variables ranging over lawless sequences).

Proof: In keeping with the informal style of sections 4.4 - 4.14, we shall prove the formulation for LS (for CL the proof is of course similar but notationally more cumbersome). Let

$$\forall\varepsilon\exists x A(\varepsilon,\varepsilon_1,\ldots,\varepsilon_n,x),$$

then there are $n_1,\ldots,n_p$ such that

$$(\varepsilon_1 \in n_1 \ \&\ldots\&\ \varepsilon_p \in n_p),$$
$$\underline{\forall}\eta_1 \in n \ldots \underline{\forall}\eta_p \in n_p \underline{\forall}\varepsilon\exists x A(\varepsilon,\eta_1,\ldots,\eta_p,x).$$

Therefore

$$\underline{\forall}\eta_1\ldots\underline{\forall}\eta_p\underline{\forall}\varepsilon\exists x(\eta_1 \in n_1 \ \ldots\ \eta_p \in n_p \rightarrow A(\varepsilon,\eta_1,\ldots,\eta_p,x))$$

and thus

$$\exists e\forall n(en\neq 0 \rightarrow \underline{\forall}\eta_1 \in k_1^{p+1}n\ldots\underline{\forall}\eta_p \in k_p^{p+1}n\underline{\forall}\varepsilon \in k_{p+1}^{p+1}n(\eta_1 \in n_1 \ \&\ldots$$
$$\ldots\&\ \eta_p \in n_p \rightarrow A(\varepsilon,\eta_1,\ldots,\eta_p,en\dot{-}1))).$$

So for some e

$$\forall n(en\neq 0 \rightarrow (\varepsilon_1 \in k_1^{p+1}n \;\&\ldots\&\; \varepsilon_p \in k_p^{p+1}n \rightarrow$$
$$\rightarrow \underline{\forall}\varepsilon \in k_{p+1}^{p+1}n\; A(\varepsilon,\varepsilon_1,\ldots,\varepsilon_p,en\dot{-}1))).$$

Now let $e[\varepsilon_1,\ldots,\varepsilon_p]m=y+1$, then for some n

$$\varepsilon_1 \in k_1^{p+1}n \;\&\ldots\&\; \varepsilon_p \in k_p^{p+1}n \;\&\; k_{p+1}^{p+1}n=m \;\&\; en=y+1,$$

and therefore $\underline{\forall}\varepsilon \in mA(\varepsilon,\varepsilon_1,\ldots,\varepsilon_p,e[\varepsilon_1,\ldots,\varepsilon_p]m\dot{-}1)$.

4.7.

Lemma. Let ϕ be a mapping from a complete, separable metric space $\Gamma=\langle V,\rho\rangle$ with basis $\langle p_n\rangle_n$ into a separable metric space Γ' with metric ρ', basis $\langle p'_n\rangle_n$. Then ϕ is *sequentially continuous*, i.e. for all $\langle x_n\rangle_n \subset V$ such that $\langle x_n\rangle_n$ converges to a limit x, $\langle \phi x_n\rangle_n$ converges to ϕx, i.e.

$$\forall k \exists n \forall m \geq n(\rho'(\phi x_m,\phi x)<2^{-k}).$$

Proof: In our proof, $\varepsilon,\varepsilon_1,\varepsilon_2,\ldots,\eta,\eta_1,\eta_2,\ldots$ are used as variables for lawless sequences. We shall say that a sequence $\alpha \in \mathcal{U}$ is a *construct* from $\varepsilon_1,\ldots,\varepsilon_i$ if for some $e \in K, \alpha=e|(\varepsilon_1,\ldots,\varepsilon_i)$. Now suppose ϕ to be a mapping depending on a non-lawlike parameter γ which is a construct from $\eta_1,\ldots,\eta_i$ (and not depending on other non-lawlike parameters), and let $\langle x_n\rangle_n \rightarrow x$, $\langle x_n\rangle_n \in V$; i.e. there is a sequence α such that $x_n=\lim_m p_{\alpha(n,m)}$, α a construct from $\varepsilon_1,\ldots,\varepsilon_j$. Without restriction we may assume $x=\lim_n p_{\beta n}$, β also a construct from $\varepsilon_1,\ldots,\varepsilon_j$. We define a new sequence $\langle x'_n\rangle_n$ such that $x'_{2n}=x$, $x'_{2n+1}=x_n$ for all n, and we consider the following tree-structure:

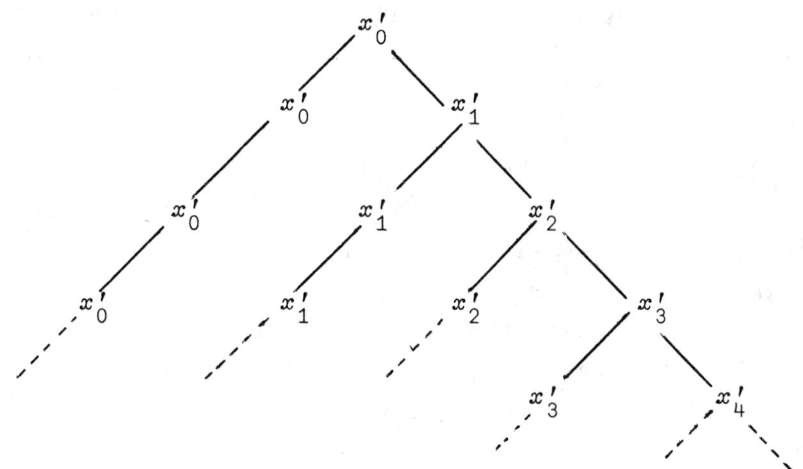

Let f be (the neighbourhood function of) a standard mapping which maps N^N into sequences of the form $0,1,2,3,\ldots,n-1,n,n,n,\ldots$; we shall assume f to be such that if $\beta\in(\overline{\lambda y.y})x$, then $f|\beta\in(\overline{\lambda y.y})x$; if $\beta\in(\overline{\lambda y.y})x*\hat{y}$ with $y\neq x$, then $f|\beta=(\overline{\lambda y.y})x*\lambda y.(x\dot{-}1)$.

$\langle x'_{(f|\beta)n}\rangle_n$ is a branch of our tree for each β, and $\lim_n x'_{(f|\beta)n}$ is a point of $\langle V,\rho\rangle$ for each β, since Γ is complete.

Therefore

(*) $\qquad \forall\beta\exists m(\rho'(\phi(\lim_n x'_{(f|\beta)n}),p'_m)<2^{-\nu})$.

Implicitly, ϕ and β contain lawless parameters $\eta_1,\ldots,\eta_i$, $\varepsilon_1,\ldots,\varepsilon_j$ (not necessarily all independent; but for simplicity in notation we shall assume them to be independent below).

It follows from (*) that

$$\underline{\forall}\varepsilon\exists m(\rho'(\phi(\lim_n x'_{(f|\varepsilon)n}),p'_m)<2^{-\nu}).$$

By the preceding lemma, abbreviating $e[\varepsilon_1,\ldots,\varepsilon_j,\eta_1,\ldots,\eta_i]$ as ξ, we have

$$\exists e\underline{\forall}\varepsilon(\rho'(\phi(\lim_m x'_{(f|\varepsilon)m}),p'_{\xi(\varepsilon)})<2^{-\nu}).$$

Let $n=(\overline{\lambda m.m})u$ be such that $\xi n\neq 0, n\neq 0$; choose ε such that $\neq(\varepsilon,\varepsilon_1,\ldots,\varepsilon_j,\eta_1,\ldots,\eta_i),\varepsilon\in n$, then

$$\underline{\forall}\varepsilon\in n(\rho'(\phi(\lim_m x'_{(f|\varepsilon)m}),p'_{\xi n\dot{-}1})<2^{-\nu}).$$

Now if $\varepsilon\in(\overline{\lambda m.m})(2z+1)*\hat{0}$, $\lim_m x'_{(f|\varepsilon)m}=x$, and if $\varepsilon\in(\overline{\lambda m.m})(2z+2)*\hat{0}$, then $\lim_m x'_{(f|\varepsilon)m}=x_z$. Therefore for all $z>u$

$\rho'(\phi x_z, \phi x) \nmid \rho'(\phi x_z, p'_{\xi n \dot- 1}) \cdot \rho'(\phi x, p'_{\xi n \dot- 1}) < 2.2^{-\nu}$.

Remark. The reason for the passage from $\langle x_n \rangle_n$ to the new sequence $\langle x'_n \rangle_n$ in the beginning of the proof is that we want to make certain that for any initial segment n of $\lambda m.m$ we can find some $\varepsilon \in n$, $\neq(\varepsilon, \varepsilon_1, \ldots, \varepsilon_j, \eta_1, \ldots, \eta_i)$ such that $\lim_n x'_{(f|\varepsilon)n} = x$. If all the x_n are different from x, this purpose is certainly not achieved for the original sequence $\langle x_n \rangle_n$ instead of $\langle x'_n \rangle_n$, since then the only sequence with limit x in the tree is $\langle x_n \rangle_n = \langle x_{(f|\lambda m.m)n} \rangle_n$; and since $f|\varepsilon = \lambda m.m$ implies $\lambda m.m = \varepsilon$, the sequence ε cannot possibly be lawless. In other respects, the proof is very similar to the proof of sequential continuity in Troelstra (1967), Theorem 2.

4.8.

Lemma. For $\mathcal{U}$ the following special instance of continuity holds

(1) $\forall \alpha \exists x \forall y \geq x A(\bar{\alpha} y, \delta) \rightarrow \exists m \forall n A(m * n, \delta)$.

Proof: Assume $\forall \alpha \exists x \forall y \geq x A(\bar{\alpha} y, \delta)$, and let $\delta = e|(\varepsilon_1, \ldots, \varepsilon_p)$. Then

$$\underline{\forall} \varepsilon_0 \exists x \forall y \geq x A(\bar{\varepsilon}_0 y, e|(\varepsilon_1, \ldots, \varepsilon_p)).$$

Hence for some $f \in K$

$$\forall m(fm \neq 0 \rightarrow \underline{\forall} \varepsilon_0 \in m \forall y \geq fm \dot- 1 \; A(\bar{\varepsilon}_0 y, e|(\varepsilon_1, \ldots, \varepsilon_p))).$$

Choose m such that $fm \neq 0$, $\mathrm{lth}(m) \geq fm \dot- 1$, then $\forall n A(m * n, \delta)$.

4.9.

Lemma. Let ϕ be a sequentially continuous mapping from a separable metric space $\Gamma = \langle V, \rho \rangle$ with basis $\langle p_n \rangle_n$ into a separable metric space $\Gamma' = \langle V', \rho' \rangle$ with basis $\langle p'_n \rangle_n$. Then ϕ is continuous, i.e.

$$\forall x \in \underline{V} \forall k \exists n \forall y \in V(\rho(x, y) < 2^{-n} \rightarrow \rho'(\phi x, \phi y) < 2^{-k}).$$

Proof: Let x be any point of Γ. We wish to construct a β such that

(1) $\forall m(\rho(p_{\beta(n,m)}, x) < 2^{-n})$

(2) $\forall m(\rho(p_m, x) < 2^{-n} \rightarrow \exists k(m = \beta(n, k)))$.

This is always possible, as follows. Obviously

$$\forall n\forall k\exists i(|r_i-\rho(p_n,x)|<2^{-k})$$

which is equivalent to

$$\forall n\forall k\exists i\{\exists k_1\exists k_2\forall m\geq k_1(|r_i-\rho(p_n,p_{\alpha m})|<2^{-k}-2^{-k_2})\},$$

hence by AC-NN* for $\mathcal{U}$ (section 4.4) there is a γ such that

$$|r_{\gamma(n,k)}-\rho(x,p_n)|<2^{-k}.$$

Now let $\langle\beta(n,m)\rangle_m$ be an enumeration of

$$U_n=\{i:\exists l(r_{\gamma(i,l)}<2^{-k}-2^{-l})\}.$$

Note that $U_n=U_n^* \equiv_{\mathrm{def}} \{i:\rho(p_i,x)<2^{-k}\}$. To see this, assume $\rho(p_i,x)<2^{-k}$; then, for some l, $\rho(p_i,x)<2^{-k}-2.2^{-l}$, therefore $r_{\gamma(i,l)}\leq\rho(p_i,x)+2^{-l}<2^{-k}-2^{-l}$, and thus $i\in U_n$. Conversely, if, for some l, $r_{\gamma(i,l)}<2^{-k}-2^{-l}$, then $\rho(x,p_i)<r_{\gamma(i,l)}+2^{-l}<2^{-k}$ so $i\in U_n^*$.

Consider all sequences $\langle p_{\beta(n,\alpha n)}\rangle_n$. All these sequences converge to x (with a prescribed rate of convergence, since $\rho(x,p_{\beta(n,m)})<2^{-n}$). We use sequential continuity (section 4.7) and obtain, for any fixed natural number ν

(3) $\quad \forall\alpha\exists k\forall n\geq k(\rho'(\phi x,p_{\beta(n,\alpha n)})<2^{-\nu}).$

Now apply the schema (1) of the preceding lemma with

$$A(m,\delta)\equiv[\rho'(\phi x,p_{\beta(\mathrm{1th}(m)\dot{-}1,(m)_{\mathrm{1th}(m)})})<2^{-\nu}]$$

where δ codes β and choice parameters of ϕ. Then by (3)

$$\exists m\forall nA(m*\hat{n},\delta);$$

hence also $\forall kA(m*\hat{k},\delta)$, i.e.

$$\forall k(\rho'(\phi x,p_{\beta(\mathrm{1th}(m),k)})<2^{-\nu}),$$

and thus

$$\rho(p_k,x)<2^{-\mathrm{1th}(m)}\rightarrow\rho(\phi x,\phi p_k)<2^{-\nu}.$$

Now consider an arbitrary y such that $\rho(y,x)<2^{-\mathrm{1th}(m)}$, $y=\lim_n p_{\gamma n}$, $\rho(p_{\gamma n},x)<2^{-\mathrm{1th}(m)}$, since we may choose $\rho(p_{\gamma n},y)<2^{-\mathrm{1th}(m)}-\rho(y,x)$ for all n. We again apply sequential continuity and obtain

$$\exists k\forall n\geq k(\rho'(\phi p_{\gamma n},\phi y)<2^{-\nu})$$

and therefore $\rho'(\phi y,\phi x)\ngtr\rho'(\phi y,\phi p_{\gamma n})+\rho'(\phi p_{\gamma n},\phi x)<2.2^{-\nu}$ (for suitable n) and thus

$$\rho(y,x)<2^{-\mathrm{1th}(m)}\rightarrow\rho'(\phi y,\phi x)<2.2^{-\nu}.$$

4.10.

Corollary. WC-N!* holds in $\mathcal{U}$ (provably in $\underset{\sim}{\mathrm{CL}}$).

Proof: Let $\forall\alpha\exists!xA(\alpha,x,\beta)$. Then A determines a mapping from Baire space into the natural numbers (a mapping with a non-lawlike parameter β). Intuitionistically, the metric of Baire space is determined by taking the sequences becoming eventually zero as basis points, and defining $\rho(\alpha,\beta)=2^{-n}$ for these points if n is the first argument for which $\alpha n\neq\beta n$, 0 otherwise. Hence $\forall\alpha\exists!x\exists y\forall\gamma\in\overline{\alpha}y\ A(\gamma,x,\beta)$.

Remark. We have actually proved a more general continuity theorem; at least it is not obvious that the continuity for mappings from complete separable metric spaces to separable metric spaces can be obtained from WC-N!* for Baire space.

4.11.

Theorem. The schema SBC*

SBC* $\quad \forall\alpha\exists xA(\overline{\alpha}x,e|\varepsilon^{u}) \to \exists f\forall\alpha A(\overline{\alpha}(f[\varepsilon^{u}](\alpha))\ e|\varepsilon^{u})$

holds for $\mathcal{U}$ (provably in $\underset{\sim}{\mathrm{CL}}$).

Proof: Let $\forall\alpha\exists xA(\overline{\alpha}x,e|\varepsilon^{u})$. Then especially

$$\forall\varepsilon^{0}(D(\varepsilon^{0},\varepsilon^{u}) \to \exists xA(\overline{\varepsilon}^{0}x,e|\varepsilon^{u})),$$

and therefore, by the lemma in section 4.6 there is an f such that

$$\forall\varepsilon^{0}(D(\varepsilon^{0},\varepsilon^{u}) \to A(\overline{\varepsilon}^{0}(f[\varepsilon^{u}](\varepsilon^{0})),e|\varepsilon^{u}).$$

Now consider any α. We may assume for f, without restriction, that $f[\varepsilon^{u}](n)\dot{-}1 \leq \mathrm{lth}(n)$. For some x $f[\varepsilon^{u}](\overline{\alpha}x)\neq 0$, hence if we choose ε^{0} such that $D(\varepsilon^{0},\varepsilon^{n})$ and $\varepsilon^{0}\in\overline{\alpha}x$, it follows that $A(\overline{\varepsilon}^{0}(f[\varepsilon^{u}](\alpha)),e|\varepsilon^{u})$. Without restriction we may assume $f[\varepsilon^{u}](\overline{\alpha}x)\dot{-}1\leq x$, therefore $A(\overline{\alpha}(f[\varepsilon^{u}](\alpha)),e|\varepsilon^{u})$.

4.12.

Corollary. The schema BI* (see section 4.2) holds in $\mathcal{U}$ (provably in $\underset{\sim}{\mathrm{CL}}$).

Proof: From SBC*; very similar to the derivation from SBC and A in 5.7.5 of Kreisel and Troelstra (1970). The role of A is taken over by CL6.

4.13.

<u>Theorem.</u> There are instances of AC-NC and WC-N in the language of $\underset{\sim}{CL}$ which can be refuted in $\underset{\sim}{CL}$.

Proof: (i) To refute AC-NC, consider the following predicate: $A(x,\alpha)\equiv\exists\varepsilon(|\varepsilon|=x \ \&\ \alpha=\varepsilon)$. Then obviously

$$\underset{\sim}{CL} \vdash \forall x\exists\alpha\ A(x,\alpha),$$

but also

$$\underset{\sim}{CL} \vdash \neg\exists\beta\forall x\ A(x,(\beta)_x)$$

because $\exists\beta\forall xA(x,(\beta)_x)$ would imply that for some e,η

$$\forall xA(x,(e|\eta)_x).$$

Suppose $|\eta|=y$; then $(e|\eta)_{y+1}=\varepsilon^{y+1}$ for some ε^{y+1}, which is impossible, because this implies that there is some $\varepsilon^0=j_i^{y+2}\varepsilon^{y+1}$ such that $D(\varepsilon^0,\eta)$, and hence

$$[(e|\eta)_{y+1}=\varepsilon^{(y+1)}] \rightarrow$$
$$\rightarrow \exists n(\varepsilon^0\in n)\ \&\ \forall\eta^0\in n(D(\eta^0,\eta)\rightarrow j_i^{y+2}(e|\eta)_{y+1}=\eta^0).$$

This is obviously false, since it would imply

$$j_i^{y+2}(e|\eta)_{y+1}=\eta_1^0=\eta_2^0$$

for $\eta_1\in n*\hat{0}$ and $\eta_2\in n*\hat{1}$.

(ii) To refute WC-N, we consider the predicate

$$B(\alpha,x)\ \equiv\ \exists e\exists\varepsilon(|\varepsilon|=x\ \&\ \alpha=e|\varepsilon)$$

Then obviously

$$\underset{\sim}{CL}\ \vdash \forall\alpha\exists xB(\alpha,x).$$

But also

$$\underset{\sim}{CL}\ \vdash \neg\forall\alpha\exists x\exists y\forall\beta\in\bar{\alpha}yB(\beta,y).$$

For consider any α, say ε^0, and assume x,y to be such that

$$\forall\beta\in\bar{\varepsilon}^0y\exists e\exists\varepsilon(|\varepsilon|=x\ \&\ \beta=e|\varepsilon).$$

Apply this to an $\eta^{x+1}\in\bar{\varepsilon}^0y$, then

$$\exists e\exists\varepsilon(|\varepsilon|=x\ \&\ \eta^{x+1}=e|\varepsilon),$$

and this is impossible by an argument similar to (i).

4.14.

<u>Theorem.</u> $\forall\alpha\exists\beta$-continuity does not even hold for the fragment of the language of $\underset{\sim}{CL}$ without the variables for u-tuples of lawless sequences.

Proof: $\forall\alpha\exists\beta\neg\exists a(\beta=a)$ obviously holds for $\mathcal{U}$, taking for β any lawless sequence; on the other hand, $\forall\alpha\exists\beta$-continuity would imply $\exists e \forall \alpha \neg \exists a(e|\alpha = a)$ which is refuted by taking for α a lawlike sequence.

4.15.

Conjecture. We have refuted AC-NC and WC-N for $\underset{\sim\sim}{\mathrm{CL}}$ using the greater expressive power of $\underset{\sim\sim}{\mathrm{CL}}$ as compared to $\underset{\sim\sim}{\mathrm{EL}}$: A and B in the proof of 4.13 could not have been defined in $\underset{\sim\sim}{\mathrm{EL}}$, since we do not have variables for u-tuples of lawless sequences there. We may conjecture however, that for formulae A of the fragment of $\underset{\sim\sim}{\mathrm{CL}}$ without variables for u-tuples of lawless sequences, WC-N and AC-NC hold for $\mathcal{U}$ (and in fact, provably in $\underset{\sim\sim}{\mathrm{CL}}$).

4.16.-28. The universes $\mathcal{U}_\alpha^*$ and the Scott-Moschovakis models.

4.16. Introductory remarks.

The principal sources in connection with the Scott-Moschovakis models are Scott (1968, 1970), Moschovakis (1973), van Dalen (1974, 1974A), Kreisel (1958), Dyson and Kreisel (1961). In Scott (1968, 1970), the so-called topological interpretation for intuitionistic predicate logic was extended to the intuitionistic theory of real numbers and real-valued functions. In Moschovakis (1973), the topological model was extended to intuitionistic analysis formulated with number-theoretic functions, so as to make comparison with existing formal systems of intuitionistic analysis possible (Kleene and Vesley (1965), Kreisel and Troelstra (1970); for a description of these systems see Chapter 5). In the final section (section 4) of Moschovakis (1973) it is shown that under the correspondence of number-theoretic functions to reals as given in Kleene and Vesley (1965), Chapter IV, the first-order part of Scott's model for the reals is contained in the model of Moschovakis (1973). The principal achievement of Scott (1970) consists of a classical proof of the validity of continuity of real-valued functions in the model; this was adapted in Moschovakis (1973) to obtain $\forall\alpha\exists!x$-continuity. In addition, Moschovakis establishes

$\forall\alpha\exists x$-continuity *without* parameters; Van Dalen (1974) provides a counterexample (formulated for an equivalent Beth model) to $\forall\alpha\exists x$-continuity *with* parameters.

Extending earlier observations (Kreisel 1958) Kreisel pointed out (Kreisel 1970, pages 137-138) that validity in (the obvious intuitionistic analogue of) Scott's model could be interpreted as intuitionistic validity for all universes $\mathcal{U}_\alpha^0 = \{e | \alpha : e \in K_r\}$ for all lawless α, where K_r denotes the class of continuous mappings carrying sequences into (canonical) real-number generators. Adapting this remark to the model of Moschovakis (1973), we can show that validity in (the intuitionistic analogue of) this model is equivalent to intuitionistic validity for all universes

$$\mathcal{U}_\alpha^* = \{e | \alpha : e \in K\}, \ \alpha \text{ lawless.}$$

The connection between validity in the topological model and validity in all $\mathcal{U}_\alpha^*$ is a special case of the equivalence between validity in Beth models and validity in intuitionistic structures discussed in section 7.12. The proof of continuity of real-valued functions, in Scott (1970) and the proofs of $\forall\alpha\exists!x$ continuity, parameter-free $\forall\alpha\exists x$-continuity in Moschovakis (1973) are classical in an essential way i.e. the argument is obviously not valid for the intuitionistic analogues described above. The counterexample to $\forall\alpha\exists x$-continuity with parameters in van Dalen (1974) is also essentially classical. It is an open question whether $\forall\alpha\exists!x$-continuity can actually be disproved (or at least its unprovability in $\underset{\sim}{\mathrm{LS}}$ shown) with respect to validity in all $\mathcal{U}_\alpha^*$.

Since we are primarily concerned in these notes with the *notions* of lawless and choice sequence, we concentrate here on the intuitionistic analogue of the topological model; our results are only partial. For the theory of the model from a classical point of view we may refer the reader to the papers mentioned above.

4.17. Description of the topological model of Moschovakis (1973).

Let us first describe the topological model of Moschovakis (1973). The valuations assign open sets in N^N to all formulae of the language $\mathcal{L}[\underset{\sim}{EL}]$ (where we shall use greek letters $\alpha,\beta,\gamma,\delta$ instead of a,b,c,d for the function variables). N is the set of natural numbers, N^N is Baire space, and K is the set of all (neighbourhood functions of) continuous operators of type $N^N \to N^N$.

We associate with each formula $A(\alpha_1,\ldots,\alpha_k,x_1,\ldots,x_n)$ containing at most $\alpha_1,\ldots,\alpha_k,x_1,\ldots,x_n$ free, for each assignment of elements $\underline{e}_i \in K$ to the α_i, and natural numbers $\underline{x}_i$ to x_i, an open subset $[\![A]\!]$ of N^N, as follows:

(i) For a prime formula $s(\alpha_1,\ldots,x_1,\ldots)=t(\alpha_1,\ldots,x_1,\ldots)$ we put $[\![s(\underline{e}_1,\ldots,\underline{x}_1,\ldots)=t(\underline{e}_1,\ldots,\underline{x}_1,\ldots)]\!] \equiv_{\text{def}}$ $\{\beta : s(\underline{e}_1|\beta,\ldots,\underline{x}_1,\ldots)=t(\underline{e}_1|\beta,\ldots,\underline{x}_1,\ldots)\}$.
Since t,s are continuous, this is an open set.

(ii) $[\![A \,\&\, B]\!] = [\![A]\!] \cap [\![B]\!]$,

(iii) $[\![A \vee B]\!] = [\![A]\!] \cup [\![B]\!]$,

(iv) $[\![A \to B]\!] = \text{Int}\{\beta : \beta \in [\![A]\!] \to \beta \in [\![B]\!]\}$ which is classically equal to $\text{Int}((N^N - [\![A]\!]) \cup [\![B]\!])$. $\text{Int}\, W \equiv_{\text{def}}$ Interior of W.

(v) $[\![\exists x A x]\!] = \bigcup_{\underline{x} \in N} [\![A\underline{x}]\!]$,

(vi) $[\![\forall x A x]\!] = \text{Int}(\bigcap_{\underline{x} \in N} [\![A\underline{x}]\!])$,

(vii) $[\![\exists \alpha A \alpha]\!] = \bigcup_{\underline{e} \in K} [\![A\underline{e}]\!]$

(viii) $[\![\forall \alpha A \alpha]\!] = \text{Int}(\bigcap_{\underline{e} \in K} [\![A\underline{e}]\!])$.

A sentence A is called *valid* if $[\![A]\!] = N^N$.
For a description of this model as an equivalent Beth model, see section 7.16.

4.18.

<u>Lemma.</u> Define $V_n = \{\beta : \beta \in n\}$. Then
$\text{Int}\{\beta : \beta \in [\![A]\!] \to \beta \in [\![B]\!]\} = \cup\{V_m : \forall n(V_{m*n} \subset [\![A]\!] \to V_{m*n} \subset [\![B]\!])\}$.

Proof: Let us put $W_1 = \text{Int}\{\beta : \beta \in [\![A]\!] \to \beta \in [\![B]\!]\}$ and let us write W_2 for the right-hand side.

Let $\beta \in W_2$. Then $\beta \in V_m$ for some m such that $\forall n(V_{m*n} \subset [\![A]\!] \to V_{m*n} \subset [\![B]\!])$. If $\beta \in [\![A]\!]$, then $\beta \in V_{m*n} \subset [\![A]\!]$ for a suitably large n, since $[\![A]\!]$ and V_m are both open; therefore $V_{m*n} \subset [\![B]\!]$, hence $\beta \in [\![B]\!]$.

Conversely, assume $\beta \in W_1$. Since W_1 is open there exists a V_m such that $\beta \in V_m \subset W_1$, and hence $\forall \gamma \in V_m(\gamma \in [\![A]\!] \to \gamma \in [\![B]\!])$, which implies $\forall n(V_{m*n} \subset [\![A]\!] \to V_{m*n} \subset [\![B]\!])$.

4.19. Reformulation of the topological interpretation.

With the help of the preceding lemma, we can now reformulate the topological interpretation as a syntactically defined mapping of formulae of EL into the set of formulae of (a sublanguage of) LS, as follows.

Suppose a fixed assignment of K-variables to function variables of EL is given. Numerical variables are assigned to themselves. (Convention: e_i is the K-variable corresponding to α_i). Let β be a lawless variable. We define the mapping Γ, $\Gamma(A)=\langle A\rangle$, by the clauses (i)*-(viii)* corresponding to (i)-(viii):

(i)* $\langle s(\alpha_1,\ldots,x_1,\ldots)=t(\alpha_1,\ldots,x_1,\ldots)\rangle \equiv_{def} s(e_1|\beta,\ldots,x_1,\ldots)=t(e_1|\beta,\ldots,x_1,\ldots)$

(ii)* $\langle A\&B\rangle \equiv_{def} \langle A\rangle \& \langle B\rangle$

(iii)* $\langle A\vee B\rangle \equiv_{def} \langle A\rangle \vee \langle B\rangle$

(iv)* $\langle A\to B\rangle \equiv_{def} \exists m(\beta\in m \& \forall n(\forall\beta\in m*n\langle A\rangle \to \forall\beta\in m*n\langle B\rangle))$

(v)* $\langle \exists x Ax\rangle \equiv_{def} \exists x\langle Ax\rangle$

(vi)* $\langle \forall x Ax\rangle \equiv_{def} \exists n(\beta\in n \wedge \forall\beta\in n\forall x\langle Ax\rangle)$

(vii)* $\langle \exists\alpha_i \forall\alpha_i\rangle \equiv_{def} \exists e_i\langle A(\alpha_i)\rangle$

(viii)* $\langle \forall\alpha_i \underline{A}\alpha_i\rangle \equiv_{def} \exists n(\beta\in n \& \forall\beta\in n\forall e_i\langle A\alpha_i\rangle)$.

All formulae $\langle A\rangle$ contain at most β as lawless variable.

4.20.

We define another mapping * as follows. The quantifiers $\forall\alpha,\exists\alpha$ are supposed to range over a universe $\mathcal{U}^*_\beta$, i.e. $\forall\alpha_i,\exists\alpha_i$ are replaced by $\forall e_i,\exists e_i$ and each occurrence of α_i in a prime formula is replaced by $e_i|\beta$, otherwise the formulae are unchanged. (Inductively:

$[t(\alpha_1,\ldots,x_1,\ldots)=s(\alpha_1,\ldots,x_1,\ldots)]^* \equiv$
$\equiv t(e_1|\beta,\ldots,x_1,\ldots)=s(e_1|\beta,\ldots,x_1,\ldots)$, $(B\&C)^*\equiv B^*\&C^*$,
$(B\vee C)^*\equiv B^*\vee C^*$, $(B\to C)^*\equiv B^*\to C^*$, $(\exists xB)^*\equiv\exists xB^*$, $(\forall xB)^*\equiv\forall xB^*$,
$(\forall\alpha_i B)^*\equiv\forall e_i B^*$, $(\exists\alpha_i B)^*\equiv\exists e_i B^*$.)

4.21.

<u>Theorem.</u> $\underset{\sim\sim}{\mathrm{LS}} \vdash A^* \leftrightarrow \langle A\rangle$.

Proof: By induction on the logical complexity of A.
(i), (ii), (iii) immediate or trivially from the induction hypothesis. (iv) $A^*\equiv(B \to C)^* \equiv (B^* \to C^*) \leftrightarrow$
$\leftrightarrow \exists m(\beta\in m \;\&\; \forall\beta\in m(\langle B\rangle \to \langle C\rangle) \leftrightarrow \exists m(\beta\in m \;\&\; \forall n(\forall\beta\in m*n\langle B\rangle \to$
$\to \forall\beta\in m*n\langle C\rangle))\equiv\langle B\to C\rangle$ (in the first equivalence the induction hypothesis was used).
(v) trivial.
(vi) $[\forall xAx]^*\equiv \forall x[Ax]^* \leftrightarrow \forall x\langle Ax\rangle \leftrightarrow \exists m(\beta\in m \;\&\; \forall\beta\in m\forall x\langle Ax\rangle) \equiv \langle\forall xAx\rangle$.
(vii) $[\exists\alpha_i\, A\alpha_i]^*\equiv \exists e_i[A\alpha_i]^* \leftrightarrow \exists e_i\langle A\alpha_i\rangle \equiv \langle\exists\alpha_i A\alpha_i\rangle$.
(viii) $[\forall\alpha_i A\alpha_i]^*\equiv\forall e_i[A\alpha_i]^* \leftrightarrow \forall e_i\langle A\alpha_i\rangle \leftrightarrow$
$\exists m[\beta\in m \;\&\; \forall\beta\in m\forall e_i\langle A\alpha_i\rangle] \equiv \langle\forall\alpha_i A\alpha_i\rangle$.

<u>Corollary.</u> Validity in the topological model, i.e. $\forall\beta A^*$, is equivalent to validity in all $\mathcal{U}^*_\beta$.

<u>Remark.</u> In establishing validity in the topological model or in all $\mathcal{U}^*_\beta$ by classical methods, we can, at least for sentences, work in $\underset{\sim}{\mathrm{IDB}}{}^c_1$, since the elimination theorem shows that the axioms for lawless sequences are compatible with *classical* logic for formulae not containing lawless variables free.

4.22. *Summary of results.*

Arguing classically, in Moschovakis (1973) the following schemas are shown to be valid in the topological model:
AC-NC* $\forall x\exists\alpha A(x,\alpha,\beta) \to \exists\gamma\forall xA(x,(\gamma)_x,\beta)$
and even stronger
RDC-C* $\forall\alpha[A(\alpha,\gamma) \to \exists\beta(B(\alpha,\beta,\gamma) \;\&\; A(\beta,\gamma))] \to$
$\to \forall\alpha[A(\alpha,\gamma) \to \exists\beta[(\beta)_0=\alpha \;\&\; \forall xB((\beta)_x,(\beta)_{Sx},\gamma)]]$;
C-N!* $\forall\alpha\exists!xA(\alpha,x,\beta) \to \exists\gamma\in K_0\forall\alpha A(\alpha,\gamma(\alpha),\beta)$ (K_0 as in 4.2);

C-N $\forall\alpha\exists xA(\alpha,x) \rightarrow \exists\gamma\in K_0\forall\alpha A(\alpha,\gamma(\alpha))$
($\forall\alpha\exists xA(\alpha,x)$ closed with respect to function variables);
BI* (as in section 4.2), and Kripke's schema:
KS $\exists\beta[A \leftrightarrow \exists x(\beta x=0)]$.

In van Dalen (1974), a counterexample to C-N* (i.e. C-N without the restriction) is given. It is plausible (from Scott (1970) and the final section of Moschovakis 1973) that here too we can extend C-N!* to the stronger theorem stating that all mappings from complete separable metric spaces into separable metric spaces are continuous; compare also the remarks in section 4.28.

By our reformulation in the theory of lawless sequences, we obtain intuitionistic proofs of the validity of AC-NC*, RDC-C*, and BI_{ar} (BI with P arithmetical, Pn, Qn without function quantifiers) for the intuitionistic analogue of the topological model.

4.23.

Theorem. AC-NC* holds in all $\mathcal{U}^*_\beta$ (provably in LS).

Assume $(\forall x\exists\alpha_1 A(x,\alpha_1,\alpha_2))^*$ which is equivalent to $\forall x\exists\alpha_1 A^*(x,\alpha_1,\alpha_2)$. Let us put $\bar{A}(x,e_1|\beta,e_2|\beta)\equiv A^*(x,\alpha_1,\alpha_2)$. By LS3 there is an m such that

$$\beta\in m \;\&\; \forall\beta\in m\forall x\exists e_1\bar{A}(x,e_1|\beta,e_2|\beta)$$

and therefore

$$\forall x\forall\beta\in m\exists e_1\bar{A}(x,e_1|\beta,e_2|\beta),$$

hence

$$\forall x\exists f\exists e\forall\beta\in m\bar{A}(x,\lambda n.e(<f(\beta)>*n)|\beta,e_2|\beta).$$

Obviously for any f,e we can find an e' such that $\forall a(\lambda n.e(<f(a)>*n)|a=e'|a)$, and therefore it follows that

$$\forall x\exists e\forall\beta\in m\bar{A}(x,e|\beta,e_2|\beta).$$

By AC-NF, there is an f such that

$$\forall x\forall\beta\in m\bar{A}(x,\lambda' n.f(\hat{x}*n)|\beta,e_2|\beta).$$

Now put

$$f'0=0,\; f'(\hat{z}*n)=f(<j_1z,j_2z>*n)$$

(and therefore $(f'|\beta)_x=\lambda' n.f(\hat{x}*n)|\beta)$, then

$$\forall\beta\in m\forall x\bar{A}(x,(f'|\beta)_x,e_2|\beta)$$

which is equivalent to

$$\forall \beta \in m(\exists \alpha_3 \forall x A(x,(\alpha_3)_x,\alpha_2))^*.$$

4.24.

Theorem. RDC-C* holds in all $\mathcal{U}^*_\beta$ (provably in $\underset{\sim}{LS}$ + RDC-F for lawlike functions).

Remark. RDC-C* implies AC-NC*, but we have given a proof of the validity of AC-NC* first because it is simpler.

Proof: Take an instance of RDC-C* (for simplicity in notation we shall disregard additional parameters); its validity in $\mathcal{U}^*_\beta$ is expressed by

$$\forall e_1[A(e_1|\beta) \to \exists e_2(B(e_1|\beta,e_2|\beta) \ \&\ A(e_2|\beta))] \to$$
$$\forall e_1[A(e_1|\beta) \to$$
$$\to \exists e_3((e_3|\beta)_0=(e_1|\beta) \ \&\ \forall n B((e_3|\beta)_n,(e_3|\beta)_{n+1}))].$$

We may replace the conclusion of the main implication by

(1) $$\forall e_1[A(e_1|\beta) \to$$
$$\to \exists e_3((e_{<0>}|\beta)=e_1|\beta \ \&\ \forall n B(e_{<n>}|\beta,e_{<n+1>}|\beta))]$$

where $e_{<x>}=\lambda' n.e(\hat{x}*n)$.

Therefore assume

$$\forall \beta \in n \forall e_1[A(e_1|\beta) \to \exists e_2(B(e_1|\beta,e_2|\beta) \ \&\ A(e_2|\beta))].$$

Then

$$\forall e_1 \forall m[\forall \beta \in n*m A(e_1|\beta) \to$$
$$\to \forall \beta \in n*m \exists e_2(B(e_1|\beta,e_2|\beta) \ \&\ A(e_2|\beta))],$$

hence

$$\forall e_1 \forall m[\forall \beta \in n*m A(e_1|\beta) \to$$
$$\to \exists e \exists f \forall \beta \in n*m(B(e_1|\beta,e_{<f(\beta)>}|\beta \ \&\ A(e_{<f(\beta)>}|\beta))].$$

Since to each e,f we can find f' such that $\forall a(f'|a=e_{<f(a)>}|a)$ it follows that

$$\forall e_1 \forall m[\forall \beta \in n*m A(e_1|\beta) \to \exists e \forall \beta \in n*m(B(e_1|\beta,e|\beta) \ \&\ A(e|\beta))].$$

Now apply RDC-F for lawlike functions:

$$\forall e[Ce \to \exists f(D(e,f) \ \&\ Cf)] \to$$
$$\to \forall e[Ce \to \exists f(f_{<0>}=e \ \&\ \forall x D(f_{<x>},f_{<Sx>}))],$$

then

$$\forall m \forall e_1[\forall \beta \in n*m A(e_1|\beta) \to$$
$$\to \exists f \forall \beta \in n*m(f_{<0>}=e_1 \ \&\ \forall x B(f_{<x>}|\beta,f_{<x+1>}|\beta))]$$

which readily is seen to imply (1) for all $\beta\epsilon n$. Since this holds for all n, we have shown validity of RDC-C* in all $\mathcal{U}^*_\beta$.

4.25.

Our next aim is to prove the validity in all $\mathcal{U}^*_\beta$ for some special instances of the bar theorem (BI_{ar}).
To see this, we first extend the mapping <> described in section 4.19 and the mapping * described in 4.20 to the language of $\underset{\sim}{IDB}_1$. We shall assume the K-variables in the formulae for which the mappings are to be defined to be disjoint from the variables e_i assigned to variables α_i; under this assumption we may assign K-variables to themselves. <> is extended by the clauses

(ix) $\langle\exists eAe\rangle \equiv \exists e\langle Ae\rangle$

(x) $\langle\forall eAe\rangle \equiv \exists n(\beta\epsilon n \;\&\; \forall\beta\epsilon n\exists e\langle Ae\rangle)$

and * by the clauses $[\exists eAe]^* \equiv \exists eA^*e, \equiv [\forall eAe]^* \equiv \forall eA^*e$; the theorem of section 4.21 remains valid.
Now we establish

4.26.

<u>Lemma.</u> Let SBC_{ar} be the schema
$\forall\alpha\exists xA(\overline{\alpha}x) \rightarrow \exists e\forall\alpha A(\overline{\alpha}(e(\alpha)))$
where An is an arithmetical formula. Then BI_{ar} is derivable from SBC_{ar} in $\underset{\sim}{IDB}_1$ (where $\underset{\sim}{IDB}_1$ is supposed to be formulated with $\alpha,\beta,\gamma,\delta$ instead of a,b,c,d as variables).

Proof: Completely analogous to Kreisel and Troelstra (1970), 5.6.1.

4.27.

<u>Theorem.</u> BI_{ar} holds for all universes $\mathcal{U}^*_\beta$ (provably in $\underset{\sim}{LS}$).

Proof: By the preceding lemma, it is sufficient to establish SBC_{ar} for all $\mathcal{U}^*_\beta$. $(\forall\alpha\exists xA(\overline{\alpha}x))^*$ becomes $\forall e\exists x\overline{A}((\overline{e|\beta})x)$ for some $\overline{A}$. Now let e_n be such that $e_n|\beta=n*(\beta)_n$, then $\forall n\exists x\overline{A}((\overline{n*(\beta)_n})x)$. Because of the result on the model of $\underset{\sim}{LS}$

constructed from a single lawless sequence (section 3.21), it follows that therefore in $\underset{\sim}{\mathrm{LS}}$: $\forall\beta\exists x\overline{A}(\overline{\beta}x)$, and hence by LS4 $\exists e\forall\beta\overline{A}(\overline{\beta}(e(\beta)))$. Since $\mathcal{U}^*_\beta$ is dense for each β, $\exists e\forall f\overline{A}((\overline{f|\beta})e(f|\beta))$; to see this, note that for each β we can find n such that $en\neq 0$, $f|\beta\in n$, and therefore if we have chosen an m such that $f|\beta\in n*m$, , $en\dot{-}1 \lhd \mathrm{th}(n*m)$, it follows, taking some $\beta_0\in n*m$, that $\overline{A}(\overline{\beta}_0(e(\beta_0)))$ and since $\overline{\beta_0}(e(\beta_0))=(\overline{f|\beta})(e(f|\beta))$, also $\overline{A}((\overline{f|\beta})e(f|\beta))$, i.e. $(\exists e\forall\alpha A(\overline{\alpha}(e(\alpha))))^*$.

4.28.

Remark. The schema of section 4.8 holds for $\mathcal{U}^*_\beta$ without parameters only:

$$\forall\alpha\exists x\forall y\geq xA(\overline{\alpha}y) \rightarrow \exists m\forall nA(m*n)$$

Inspection of the proof in section 4.7 shows that the proof of sequential continuity in a point x breaks down for the $\mathcal{U}^*_\beta$, because in general we cannot introduce a lawless variable ranging over lawless sequences distinct from the ones on which x is dependent; the argument is valid however for lawlike x which do not depend on a lawless parameter. Thus relative to $\mathcal{U}^*_\beta$, we obtain sequential continuity in lawlike points; and using (1) in section 4.8 without parameters, the argument of section 4.9 also carries through for lawlike points, and thus we have as a partial result:

Theorem. Relative to the universes $\mathcal{U}^*_\beta$, mappings from complete separable metric spaces into separable metric spaces are continuous in *lawlike* points. It is an open problem whether they are continuous at all points.

5
CS: AN AXIOMATIC SYSTEM EXPRESSING THE CONTINUITY OF LOGICAL OPERATIONS

5.1.

In the preceding section we described models for intuitionistic analysis constructed from lawless sequences, which satisfied some of the typical continuity axioms playing a rôle in the literature on intuitionistic mathematics. The results show how far we got (up till now) starting from the lawless sequences. Now we shall start at the other end, and write down axioms which include the (positive) properties of the models of the preceding section, and more, e.g. the axioms of the system FIM (= the system in Kleene and Vesley (1965); FIM = 'foundations of intuitionistic mathematics', a reference to the title of Kleene and Vesley's monograph); FIM includes very strong ($\forall\alpha\exists\beta$-)continuity axioms.

Naturally, since we now have simply started at the other end, adding the strongest axioms used in practice, we need a consistency proof. In fact, we achieve more: an elimination procedure σ is described similar to the mapping τ in Chapter 3 which achieved elimination of lawless sequences and permitted us to treat quantification over lawless sequences as a 'figure of speech'; here, σ permits us to regard quantification over choice sequences as a 'figure of speech'. Expressed differently, we may regard the choice quantifiers $\forall\alpha, \exists\alpha$ as a special sort of quantification over functions which forces in quantifier combinations $\forall\alpha\exists x, \forall\alpha\exists\beta$ the x or β to depend continuously on α (i.e. the instantiations of existential statements are supposed to depend continuously on (choice-)function parameters).

Appendix C discusses models of intrinsic interest for CS. But even without having a model (i.e. a concept of choice sequence for which the axioms of CS can be derived), the treatment of choice sequences as a 'figure of speech' and especially the details of the elimination mapping have various interesting applications such as the reduction of theories (of choice sequences) including continuity and bar induction to theories of non-iterated generalized inductive definitions and the fact that an intuitionistic form of König's lemma (a weak form of the fan theorem) is conservative over intuitionistic arithmetic[†] (Troelstra (1974); see section 5.5), a result which in its statement does not refer to choice sequences at all. Another possible type of application, which has been insufficiently explored up till now, is the use of σ to obtain constructive (in the narrow sense of e.g. Bishop (1967), i.e. not involving choice sequences and/or continuity axioms) equivalents of classical theorems - see our remarks in section 6.12.

We insert here a remark on the proof of the result on König's lemma mentioned above. The idea of the proof is to *define* a class of neighbourhood functions (corresponding to the class K for CS) with certain properties built-in which are in the familiar ('classical') theory proved by non-intuitionistic means, and by axioms specific to choice sequences in the intuitionistic case. It should be noted here that straightforward 'model-building' for a class of sequences satisfying the fan theorem naturally leads to the arithmetic functions, which can be treated in the first level of the ramified hierarchy, with proof-theoretic ordinal $\varepsilon_{\varepsilon_0}$ (cf. Kleene and Vesley (1965), page 116, lemma 9.12^{c} and Kreisel (1966)).

The system in Kreisel (1963) (S for short) was a precursor of CS, but the crucial axiom (corresponding to the principle of analytic data A of CS in section 5.2) stating that every property asserted of a choice sequence should hold for all choice sequences in a certain lawlike spread containing α, was shown (in Troelstra 1968) to be incompat-

---

[†]modulo the result of Goodman (1968) that EL_1 is conservative over HA. Another proof of this result is contained in Minc (1975).

ible with closure of choice sequences under continuous operations (a theorem of **S**). It was thought that **S** was justified for a concept of choice sequence more or less corresponding to Brouwer's notion of choice sequence, and the aim was to put down enough axioms for this concept for an elimination theorem to be proved. Replacing **S** by **CS** made the elimination possible, but completely severed the connection with Brouwer's concept, so there was a new problem, namely to find a notion satisfying the axioms of **CS**. An attempt at such a notion is discussed at length in Appendix C; the attempt is not very satisfactory (not surprising, in view of the fact that the final form of **CS** was determined by formal considerations: the possibility of an elimination), mainly because the notion proposed is complicated and not very natural. It suggests nevertheless further possible developments of independent interest, which was the main reason for including Appendix C. For more details on the history of **CS**, see Appendix A, especially section A5. Principal sources for the metamathematics of **CS** are Kreisel and Troelstra (1970) and Troelstra (1971, 1974, 1974B) which the reader may consult for details.

5.2. *Description of* **CS**.

Let $\mathbf{IDB}_1$ be defined as in section 3.1. **CS** is an extension, obtained by addition of variables $(\alpha,\beta,\gamma,\delta,\ldots)$ for choice sequences, a choice-abstraction operator λ'', correspondingly extended rules of term and functor formation (cf. Kreisel and Troelstra (1970), 5.2.1), notably:

> If t is a term, then $\lambda''x.t$ is a choice functor; if ϕ is a K-functor, ϕ' a choice-functor, then $\phi(\phi')$ is a term and $\phi|\phi'$ a choice-functor.

Axioms and rules for choice quantifiers are added, and a rule of λ''-conversion,

λ''-CON $\quad (\lambda''x.t[x])t'=t[t']$

and defining axioms for $e(\alpha), e|\alpha$:

$$e(\alpha)=x \ \& \ e\bar{\alpha}y=Sz \rightarrow z=x,$$

$$(e|\alpha)(x)=(\lambda'n.e(\hat{x}*n))(\alpha).$$

The rules and axiom schemas of $\underset{\sim}{\mathrm{IDB}}_1$ are extended to this language - also for instances with free choice parameters, except for AC-NF where no free choice parameters are admitted.

Finally we add the following specific axioms for choice sequences

A $\quad A\alpha \to \exists e[\exists\beta(\alpha=e|\beta)\ \&\ \forall\gamma A(e|\gamma)]$

BC-F $\quad \forall\alpha\exists a A(\alpha,a) \to \exists e\exists b\forall\alpha A(\alpha,(b)_{e(\alpha)})$

BC-C $\quad \forall\alpha\exists\beta A(\alpha,\beta) \to \exists e\forall\alpha A(\alpha,e|\alpha).$

The schema A is called the 'principle of analytic data'. A is equivalent to

A' $\quad \forall e(\forall\alpha A(e|\alpha) \to \forall\alpha B(e|\alpha)) \to \forall\alpha(A\alpha \to B\alpha)$

(cf. Kreisel and Troelstra (1970), 5.3.2).

BC-F implies as variants and special cases:

(1) $\quad \forall\alpha\exists x A(\alpha,x) \to \exists e\forall n(en\neq 0 \to \forall\alpha A(n|\alpha,en\dot{-}1)),$

(2) $\quad \forall\alpha\exists a A(\alpha,a) \to \exists e\exists b\forall n(en\neq 0 \to \forall\alpha A(n|\alpha,\lambda m.b(\langle en\dot{-}1\rangle*m))),$

(3) $\quad \forall\alpha\exists f A(\alpha,a) \to \exists e\exists f\forall n(en\neq 0 \to \forall\alpha A(n|\alpha,\lambda'm.f(\langle en\dot{-}1\rangle*m))).$

Here $n|\alpha$ abbreviates $\lambda'm.k(n,m)|\alpha$ (cf. section 3.1).
Note that in A' and (1), (2), (3) the main implication can be strengthened to equivalence.

5.3. *Definition of the translation* σ.

We now describe the definition of the mapping of formulae of $\underset{\sim}{\mathrm{CS}}$ closed with respect to choice sequences into formulae of $\underset{\sim}{\mathrm{IDB}}_1$ (this mapping is called τ in Kreisel and Troelstra (1970); we have renamed it σ here so as to avoid confusion with the mapping τ used for $\underset{\sim}{\mathrm{LS}}$ in Chapter 3). We shall in our description assume K to be completely eliminated from $\underset{\sim}{\mathrm{CS}}$ and $\underset{\sim}{\mathrm{IDB}}_1$ using K-variables instead.

We define the *degree* $\rho(A)$ of a formula A closed with respect to choice variables as $\omega.\rho'(A)+\rho''(A)$, where $\rho'(A)$ indicates the number of occurrences of $\vee$ within the scope of a choice quantifier, and $\rho''(A)$ as the number of occurrences of logical operators within the scope of a choice quantifier. (If we treat $\vee$ as defined, which is possible (cf. e.g. Troelstra (1973), 1.3.7) then we can drop $\omega.\rho'(A)$.)

We first define the auxiliary mapping $\mapsto$ for formulae with a choice quantifier in front and without choice parameters.

(i) $\forall\alpha(t[\alpha]=s[\alpha]) \mapsto \forall a(t^N[a]=s^N[a])$
(where t^N is obtained from t by replacing occurrences of choice-abstraction $\lambda''x.$ by lawlike-abstraction $\lambda x.$)

(ii) $\forall\alpha(A\alpha \ \& \ B\alpha) \mapsto \forall\alpha A\alpha \ \& \ \forall\alpha B\alpha$

(iii) $\forall\alpha(A\alpha \vee B\alpha) \mapsto \forall\alpha\exists x[(x=0 \to A\alpha) \ \& \ (x\neq 0 \to B\alpha)]$

(iv) $\forall\alpha\forall xA(\alpha,x) \mapsto \forall x\forall\alpha A(\alpha,x)$
$\forall\alpha\forall aA(\alpha,a) \mapsto \forall a\forall\alpha A(\alpha,a)$
$\forall\alpha\forall eA(\alpha,e) \mapsto \forall e\forall\alpha A(\alpha,e)$

(v) $\forall\alpha\forall\beta A(\alpha,\beta) \mapsto \forall e\forall f\forall\alpha A(e|\alpha,f|\alpha)$

(vi) $\forall\alpha(A\alpha \to B\alpha) \mapsto \forall e(\forall\alpha A(e|\alpha) \to \forall\alpha B(e|\alpha))$

(vii) $\forall\alpha\exists xA(\alpha,x) \mapsto \exists e\forall n(en\neq 0 \to \forall\alpha A(n|\alpha,en\dot{-}1))$

(viii) $\forall\alpha\exists aA(\alpha,a) \mapsto \exists e\exists b\forall n(en\neq 0 \to \forall\alpha A(n|\alpha,\lambda'm.b(\langle en\dot{-}1\rangle * m)))$

(ix) $\forall\alpha\exists e'A(\alpha,e') \mapsto \exists e\exists f\forall n(en\neq 0 \to \forall\alpha A(n|\alpha,\lambda'm.f(\langle en\dot{-}1\rangle * m)))$

(x) $\forall\alpha\exists\beta A(\alpha,\beta) \mapsto \exists e\forall\alpha A(\alpha,e|\alpha)$

(xi) $\exists\alpha A\alpha \mapsto \exists aA(\lambda''x.ax)$.

Of course the variables a in (i), x in (iii), e,f,n in (v)-(x) are assumed to be different from the variables in A,B.

Note that every application of $\mapsto$ to a suitable formula F of A strictly lowers $\rho(A)$, and therefore the process of repeated application of $\mapsto$ terminates; we call the result $\sigma(A)$.

5.4. *The elimination theorem.*

For the translation σ one can prove a theorem similar to that in section 3.3:

(i) $\sigma(A)\equiv A$ for formulae of $\mathbf{IDB}_1$,

(ii) $\mathbf{CS} \vdash A \leftrightarrow \sigma(A)$,

(iii) $\mathbf{CS} \vdash A \Leftrightarrow \mathbf{IDB}_1 \vdash \sigma(A)$.

The proof is given in full detail in Kreisel and Troelstra (1970), section 7.

(i) is obvious from the definition

(ii) is readily verified. Note that

(iv) A' and (1), (2), (3), BC-C in section 5.2 can be

strengthened to equivalences,
(v) $\exists\alpha A\alpha \leftrightarrow \exists a A(\lambda''x.ax)$ is a consequence of BC-C applied to $\forall\beta\exists\alpha A\alpha$, since this yields $\exists e\forall\alpha A(e|\alpha)$, hence $\exists eA(e|\lambda''x.0)$,
(vi) $\forall\alpha(t[\alpha]=s[\alpha]) \leftrightarrow \forall a(t^N[a]=s^N[a])$ follows from the fact that terms are continuous in their function parameters; with the help of (iv), (v), (vi) one then easily sees that $\mapsto$ replaces each formula by a provably equivalent one.

The proof of (iii) is based on three lemmas:

<u>Lemma A.</u> $\forall a(t^N[a]=s^N[a]) \rightarrow (\ulcorner\forall\alpha A(t[\alpha],\alpha)\urcorner \leftrightarrow \ulcorner\forall\alpha A(s[\alpha],\alpha)\urcorner)$
$\forall a(\phi^N[a]=\psi^N[a]) \rightarrow (\ulcorner\forall\alpha A(\phi[\alpha],\alpha)\urcorner \leftrightarrow \ulcorner\forall\alpha A(\psi[\alpha],\alpha)\urcorner)$.

<u>Lemma B.</u> $\forall e(\ulcorner\forall\alpha A\alpha\urcorner \rightarrow \ulcorner\forall\alpha A(e|\alpha)\urcorner)$.

<u>Lemma C.</u> $\forall e(\ulcorner\forall\alpha A(e(\alpha)\ \alpha)\urcorner \leftrightarrow \forall n(en\neq 0\rightarrow \ulcorner\forall\alpha A(en\dot{-}1,n|\alpha)\urcorner))$.

In the statement of these lemmas, $t^N[a], s^N[a]$ are obtained from $t[a]$, $s[a]$ by replacement of λ'' by λ; $\ulcorner A\urcorner$ indicates $\sigma(A)$.

Lemmas A,B,C are proved by induction on the degree of $\forall\alpha A$; and with the help of these lemmas one then proves (iii) by induction on the length of deductions in $\underset{\sim}{\mathrm{CS}}$.

5.5. *The elimination theorem in a more general setting; some applications.*

Let now $\underset{\sim}{\mathrm{H}}$ be a system in the language of $\underset{\sim}{\mathrm{IDB}}_1$ with the constant K omitted (but K-variables retained), axiomatized by the axioms and schemas of $\underset{\sim}{\mathrm{EL}}$, AC-NF, and certain postulates for K (expressed via K-variables of course).

$\underset{\sim}{\mathrm{H}}^*$ is the same system as $\underset{\sim}{\mathrm{H}}$ but with another set of variables for lawlike functions; we use greek letters $\alpha,\beta,\gamma,\delta$ instead of the a,b,c,d used to denote lawlike variables in $\underset{\sim}{\mathrm{H}}$.

$\underset{\sim}{\mathrm{H}}$ can be extended to a theory $\underset{\sim}{\mathrm{CS}}_{\mathrm{H}}$ for choice sequences, in the language of $\underset{\sim}{\mathrm{CS}}$ (except for the omission of K), by addition of three postulates:

A1 $\forall\alpha\exists x(e(\overline{\alpha}x)\neq 0)$,

A3 $\forall\alpha\exists\beta A(\alpha,\beta) \rightarrow \exists e\forall\alpha A(\alpha,e|\alpha)$,

A4 $\forall\alpha[A\alpha \rightarrow B\alpha] \leftrightarrow \forall e[\forall\alpha A(e|\alpha) \rightarrow \forall\alpha B(e|\alpha)]$.

For certain applications below we might still add

A2 $\forall\alpha\exists a A(\alpha,a) \rightarrow \exists e\exists b\forall\alpha A(\alpha,(b)_{e(\alpha)})$

so as to increase the similarity with $\underset{\sim}{\mathrm{CS}}$, but A2 is not necessary for these applications, and has to be deleted for another one.

Now let $\mathfrak{A}$ be a set of additional postulates (formulated in the language of $\underset{\sim}{\mathrm{H}}$), and $\mathfrak{A}^*$ their transcription into the language of $\underset{\sim}{\mathrm{H}}^*$. We shall assume

$$\mathrm{H}^* + \mathfrak{A}^* \subset \underset{\sim}{\mathrm{CS}}_{\mathrm{H}}.$$

Provided the postulates for K contained in $\underset{\sim}{\mathrm{H}}$ permit us to prove certain closure conditions on K needed in the proof of (iii) of the elimination theorem (for a specification of these closure conditions see Troelstra (1974)), we can establish

$$\underset{\sim}{\mathrm{CS}}_{\mathrm{H}} \vdash A \Leftrightarrow \underset{\sim}{\mathrm{H}} \vdash \sigma(A)$$

for all sentences A of $\underset{\sim}{\mathrm{CS}}_{\mathrm{H}}$ not containing quantifiers $\exists a, \exists e$ within the scope of choice quantifiers $\forall\alpha$ (note that because of this restriction we do not need a transformation of the form $\forall\alpha\exists a A(\alpha,a) \mapsto \exists e\exists b\forall n(en\neq 0 \rightarrow \forall\alpha A(n|\alpha,\lambda' m.b(\langle en\dot{-}1\rangle * m)))$.

Let A^* denote the rewriting of an A in the language of $\underset{\sim}{\mathrm{H}}$ into the language of H^*, and assume

$$\underset{\sim}{\mathrm{H}} + \mathfrak{A} \vdash A.$$

Then also

$$\underset{\sim}{\mathrm{H}}^* + \mathfrak{A}^* \vdash A^*,$$

hence

$$\underset{\sim}{\mathrm{CS}}_{\mathrm{H}} \vdash A^*,$$

and therefore

$$\underset{\sim}{\mathrm{H}} \vdash \sigma(A^*).$$

If A is arithmetical, $A^* \equiv A \equiv \sigma(A^*)$, and therefore $\underset{\sim}{\mathrm{H}} + \mathfrak{A}$ is then conservative over $\underset{\sim}{\mathrm{H}}$.

If A does not contain $\vee, \exists$ within the scope of a universal function quantifier, one easily verifies $\underset{\sim}{\mathrm{H}} \vdash \sigma(A^*) \leftrightarrow A$ (by induction on the logical complexity of A). The only case which is not immediately obvious is that where $A \equiv \forall\alpha(B\alpha \rightarrow C\alpha)$; the auxiliary mapping $\mapsto$ replaces this by $\forall e(\forall\alpha B(e|\alpha) \rightarrow \forall\alpha C(e|\alpha))$, which is equivalent to $\forall\alpha(B\alpha \rightarrow C\alpha)$, since for any given α, we may take e to be such that $\forall\beta(e|\beta=\alpha)$.

Now we can not only retrieve by a few additional considerations the original elimination theorem, but we also obtain:

Corollary A. $\underset{\sim}{H}$ + $CONT_1$ is conservative over $\underset{\sim}{H}$ with respect to arithmetical sentences ($\underset{\sim}{H}$ a definitional extension of $\underset{\sim}{EL}_1$). Here $CONT_1$ is

$$\forall a\exists bA(a,b) \rightarrow \exists e\forall aA(a,e|a).$$

Proof: Take as postulate for K in $\underset{\sim}{H}$ the definition as a set of neighbourhood functions:

$$\forall a\exists x(e(\overline{a}x)\neq 0) \ \& \ \forall nm(en\neq 0 \rightarrow en=e(n*m)).$$

Corollary B. $\underset{\sim}{H}$ + $CONT_1$ + FAN is conservative over $\underset{\sim}{H}$ with respect to arithmetical sentences ($\underset{\sim}{H}$ a definitional extension of $\underset{\sim}{EL}_1$). Here FAN is

$$\forall a\leq\lambda x.1\exists xA(\overline{a}x) \rightarrow \exists z\forall a\leq\lambda x.1\exists y\forall b\leq\lambda x.1(\overline{a}z=\overline{b}z \rightarrow A(b,y))$$

where $b\leq c \equiv_{def} \forall x(bx\leq cx)$.

Proof: Take as postulate for K in $\underset{\sim}{H}$ the definition of K as a set of neighbourhood functions, uniformly continuous on *bounded* sets of functions:

$$\forall c\exists z\forall b\leq c(e\overline{b}z\neq 0) \ \& \ \forall nm(en\neq 0 \rightarrow en=e(n*m))$$

Note that, *classically*, this condition on K-functions is equivalent to the one used for corollary A; but the equivalence is not provable in $\underset{\sim}{H}$. It is easy to see that the enforced uniform continuity on *bounded* sets of functions forces the σ-translation of instances of FAN (written with $\alpha,\beta,\gamma,\delta$ instead of a,b,c,d) to become provable. For more details of the proofs, see Troelstra (1974).

Corollary C. Let $\underset{\sim}{FIM}$ be (essentially) the system of Kleene and Vesley (1965), i.e. axiomatized as $\underset{\sim}{EL}$ (written with variables $\alpha,\beta,\gamma,\delta$) + AC-NC* + BI* + $CONT_1$, where AC-NC* is defined as in section 4.2, and BI* indicates the schema of bar-induction with parameters. $\underset{\sim}{CS}$ is a conservative extension of $\underset{\sim}{FIM}$ (cf. Troelstra 1971).

Proof: $(\underset{\sim}{IDB}_1)^*$, i.e. $\underset{\sim}{IDB}_1$ with variables a,b,c,d replaced

by $\alpha,\beta,\gamma,\delta$, is a subsystem of $\underset{\sim}{FIM}$, if we define K simply as the set of neighbourhood functions (cf. Kreisel and Troelstra (1970), 6.3).
If we take any formula A in the language of $\underset{\sim}{FIM}$, and we apply the translation σ, then

$$\underset{\sim}{CS} \vdash A \Leftrightarrow \underset{\sim}{IDB}_1 \vdash \sigma(A).$$

Note that $(\sigma(A))^* \leftrightarrow A$ can be established in $\underset{\sim}{FIM}$, since A does not contain quantifiers $\exists e$, $\exists a$, so there is no need to use clauses (viii), (ix) in the definition of $\sigma(A)$, and correspondingly $(\sigma(A))^* \leftrightarrow A$ can be established without appeal to axiom schema BC-F (cf. section 5.4 (ii)). As noted above, $(\underset{\sim}{IDB}_1)^* \subset \underset{\sim}{FIM}$, and therefore $\underset{\sim}{FIM} \vdash (\sigma(A))^*$, hence also $\underset{\sim}{FIM} \vdash A$.

5.6. *Markov's principle and Church's thesis in* $\underset{\sim}{CS}$.

It is obvious that choice sequences satisfying $\underset{\sim}{CS}$ cannot all be recursive, since

CT_c $\quad \forall\alpha\exists x\forall y\exists z[Txyz \,\&\, Uz = \alpha y]$

would in conjunction with $\forall\alpha\exists x$-continuity require the x to depend on an initial segment of α which is obviously false.

On the other hand, $\underset{\sim}{IDB}_1$ + CT (CT as in section 1.3) is consistent, as can be seen by realizability methods (Kreisel and Troelstra (1970) 3.7; Troelstra (1973) 3.2.30).
In $\underset{\sim}{CS}$ we also have

(1) $\quad \forall\alpha\neg\neg\exists a[\alpha = a]$.

This is seen by applying the schema A to $\neg\exists a[\alpha = a]$; then it follows for some e that $\exists\beta(\alpha = e|\beta)$ and $\forall\gamma\neg\exists a[e|\gamma = a]$. The latter assertion is obviously false (take $\gamma = \lambda x.0$), hence $\neg\neg\exists a[\alpha = a]$.

Another proof proceeds as follows: Assume $\neg\exists a[\alpha = a]$, then $\exists\alpha\neg\exists a[\alpha = a]$, so $\forall\beta\exists\alpha\neg\exists a[\alpha = a]$; BC-C yields $\exists e\forall\beta\neg\exists a[e|\beta = a]$ which again is contradictory (i.e. (1) follows either from A or from BC-C).

On the other hand, with BC-F one readily shows

$$\neg\forall\alpha\exists a(\alpha = a).$$

We now turn to Markov's principle for choice sequences.
We state first the intuitionistic form of the well-known

example of the recursively well-founded, but not well-founded primitive recursive binary tree (Kleene (1952); Kleene and Vesley (1965), lemma 9.8; Troelstra (1973) 2.6.9).

Lemma. There is a primitive recursive predicate An, such that $\{n: An\}$ is a set of finite sequences of natural numbers which forms a tree (i.e. $A(\langle\rangle), A(n*m) \to An$) satisfying

(2) $\forall\alpha\in\mathrm{Rec}\,\exists x\neg A(\overline{\alpha}x)$

(3) $\forall x\exists n(\mathrm{lth}(n) = x \;\&\; \forall y<x((n)_x \leq 1) \;\&\; An)$.

Here 'Rec' indicates the class of all total recursive functions. As follows from the lemma, $\forall\alpha \leq \lambda x.1\, \exists x\neg A(\overline{\alpha}x)$ does not hold if the $\alpha \leq \lambda x.1$ satisfy the fan theorem.

Lemma. Let H be a system in the language of CS, but instead of the axioms A, BC-F, BC-C we assume (1) and the fan theorem in the form

(4) $\forall\alpha\leq\lambda x.1\exists xA(\overline{\alpha}x) \to \exists z\forall\alpha\leq\lambda x.1\exists x\leq zA(\overline{\alpha}x)$

(A primitive recursive). Then the following weak form of Markov's principle for choice sequences

(5) $\forall\alpha\leq\lambda x.1\neg\neg\exists xA(\overline{\alpha}x) \to \neg\neg\forall\alpha\leq x.1\exists xA(\overline{\alpha}x)$

(A primitive recursive) implies $\neg$CT.

Proof: Let A be as in the preceding lemma; then CT, (1) and (2) imply $\forall\alpha\neg\neg\exists x\neg A(\overline{\alpha}x)$, and with (5) $\neg\neg\forall\alpha\leq\lambda x.1\exists x\neg A(\overline{\alpha}x)$; but with (3), (4) $\neg\forall\alpha\leq\lambda x.1\exists x\neg A(\overline{\alpha}x)$, i.e. (5) $\to\neg$CT in H.

The interest of (5) stems from its connection with completeness problems for intuitionistic predicate logic; see section 7.11.

For CS itself we have the following theorem

Theorem. (Proof in Troelstra (1974A)).

M $\quad\forall a[\neg\neg\exists x(ax = 0) \to \exists x(ax = 0)]$,

M_c $\quad\forall\alpha[\neg\neg\exists x(\alpha x = 0) \to \exists x(\alpha x = 0)]$,

$K_0 a \equiv_{\mathrm{def}} \forall b\exists x((a(\overline{b}x) \neq 0) \;\&\; \forall nm(an\neq 0 \to an = a(n*m)))$.

Then for the translation :

$$\mathrm{CS} + M_c \vdash A \Leftrightarrow \mathrm{IDB}_1 + (K_0 = K) + M \vdash \sigma(A),$$

where $K_0 = K$ abbreviates $\forall a\in K_0\exists e(a = e)$. ($K_0 = K$ expresses the fact that all neighbourhood functions on *lawlike* sequences are inductively defined.)

5.7. *Realizability.*

S. C. Kleene introduced (Kleene 1957, 1965, 1965A, 1968, 1969 ; Kleene and Vesley 1965) another interpretation of intuitionistic analysis (the system $\underset{\sim}{\text{FIM}}$ of Corollary C in section 5.5) which may also be regarded as an elimination of choice sequences in a weaker sense. Kleene's notion, realizability (by functions) does not permit us to regard $\forall\alpha,\exists\alpha$ as new quantifiers with special properties in addition to the ordinary quantifications $\forall a,\exists a$, but it does provide a reinterpretation of functional quantification validating the continuity axioms. The interpretation does not offer new insights into the concept of choice sequence, but is technically very useful. A concise description, together with further references, can be found in Troelstra (1973), section 3.3, applications in 3.6.17-20. We summarize here the basic result.

Definitions.

(i) A formula in the language of $\underset{\sim}{\text{EL}}$ is said to be *almost negative*, if it does not contain $\vee$, and $\exists$ only in front of prime formulae.

(ii) Γ_0 consists of the formulae A such that in all subformulae of A of the form $B \to C$, $\neg B$ B is an almost negative formula preceded by a string of existential quantifiers.

(iii) Let us put

$$\alpha|\beta \simeq \gamma \equiv_{\text{def}} \forall x\exists y(\overline{\alpha}(\hat{x}*\overline{\beta}y) = \gamma x+1 \;\&\; \forall z<y(\overline{\alpha}(\hat{x}*\overline{\beta}z) = 0)).$$

Then GC (generalized continuity) is the following schema

GC $\quad \forall\alpha[A\alpha \to B(\alpha,\beta)] \to \exists\gamma\forall\alpha[A\alpha \to \exists\delta(\gamma|\alpha \simeq \delta \;\&\; B(\alpha,\delta))]$

where A almost negative, β not free in A.

The basic result is now

Theorem (Troelstra 1973, 3.3.11, 3.6.18(i)). Let $\underset{\sim}{\text{B}}$ be the system $\underset{\sim}{\text{FIM}}$ minus CONT_1, and let A' express 'A is realizable by functions' (i.e. $\exists\alpha(\alpha\, \underset{\sim}{\text{r}}^1 A)$ in the notation of Troelstra (1973)).

Then

(i) $\underset{\sim}{\text{B}} + \text{GC} \vdash A \leftrightarrow A'$

(ii) $\underset{\sim}{\text{B}} + \text{GC} \vdash A \Leftrightarrow \underset{\sim}{\text{B}} \vdash A'$

(iii) $(\underset{\sim}{\mathrm{B}} + \mathrm{GC}) \cap \Gamma_0 = \underset{\sim}{\mathrm{B}} \cap \Gamma_0$, i.e. $\underset{\sim}{\mathrm{B}} + \mathrm{GC}$ is conservative over $\underset{\sim}{\mathrm{B}}$ with respect to formulae of Γ_0

(iv) (i) - (iii) hold if we drop BI* from $\underset{\sim}{\mathrm{B}}$, or if we replace it by the fan theorem.

There is yet another interpretation, special realizability; this validates CONT_1 and IP: $(\neg A \to \exists\alpha B) \to \exists\alpha(\neg A \to B)$ (α not free in A). (See Troelstra (1973), section 3.4.)

6

CONTINUITY POSTULATES IN INTUITIONISTIC MATHEMATICS

6.1.

The present section is devoted to the presentation of some examples of the use of continuity postulates in intuitionistic mathematics. For the proofs we either give a reference to the literature or give an informal presentation. In all examples the proof depends on constructing a suitable tree of sequences of natural numbers so as to represent a certain set (= species) as an image of this tree, with the result that the assumptions of the theorem to be proved can be translated into statements of $\forall\alpha\exists x$-form, to which then C-N* for choice sequences or the fan theorem can be applied.

The examples by no means require all of $\underset{\sim}{\mathrm{CS}}$; for applications I, VII we need the fan theorem; all other applications need at most $\forall\alpha\exists x$-continuity; in fact inspection of the proofs of III, IV, X shows that they use only the following consequence of continuity:

$$\forall\alpha\exists x\forall y\geq xA(\overline{\alpha}y,\beta)\rightarrow\exists m\forall nA(m*n,\beta).$$

Since by section 4.8 this schema holds for $\mathcal{U}$, we know that III, IV, X below also hold for $\mathcal{U}$.

As to $\mathcal{U}^*_\beta$, see the remark in section 4.28.

The end of this chapter (6.12) briefly discusses the possible use of the elimination mapping for choice sequences to obtain versions of classical theorems which are constructive in the narrow sense (i.e. do not refer to choice sequences; for example theorems of 'lawlike' analysis, as in Bishop (1967)).

6.2. *Application I: uniform continuity of real-valued functions on the closed interval.*

This is the 'classic' application due to Brouwer (see e.g. Brouwer 1924) and can be stated as: every real-valued function on a closed interval is uniformly continuous. The proof uses the fan theorem; see e.g. Troelstra (1969).

6.3. *Application II: sequential continuity of functions defined on complete metric spaces.*

In full: a sequentially continuous mapping f from a complete metric space to a separable metric space is sequentially continuous (where a mapping f is said to be sequentially continuous if for any sequence $\langle x_n \rangle_n$ converging to a limit x, $\langle fx_n \rangle_n$ converges to fx). For the proof see Troelstra (1969); the proof is very similar to the one used in section 4.7. Only there we assumed the domain space to be separable as well, so as to make the formalizability in the theory $\underset{\sim}{\text{CL}}$ obvious.

It is not clear how to formalize the idea of a complete metric space in its generality in the language of $\underset{\sim}{\text{EL}}_1$, although we can show that we can define within $\underset{\sim}{\text{EL}}_1$ complete metric spaces which cannot be shown to be separable; on the other hand, no mathematically interesting example of such a space is known. As our example we may take the space $\Gamma = \langle V, \rho \rangle$ defined as follows. Let $V_0 = \{r_i : A(i)\}$ be a set of rationals (r_i referring to a fixed standard enumeration $\langle r_n \rangle_n$ of the rationals) where for each i $A(i)$ is an unsolved problem. Let V be the species of equivalence classes of all Cauchy-sequences of elements of V_0 (relative to the usual metric of the reals). Then $\langle V, \rho \rangle$ (ρ the metric induced by the metric of the real line) is a subspace of the real line which is complete but which cannot be shown to be separable (V_0 might be empty).

6.4. *Application III: Sequential continuity implies continuity.*

In full: a sequentially continuous mapping from a separable metric space to a metric space is sequentially continuous. Proof e.g. in Troelstra (1969), 13.1.3; the method is similar to the one used in section 4.9, and moreover resembles the proof of the next application.

6.5. *Application IV: Equivalence of two notions of differentiability.*

Consider the following two notions of differentiability at point x of the real line $\underset{\sim}{R}$ (apartness, #, being defined as usual in the intuitionistic literature):

(A) f has a derivative f' at x when

$$\forall\varepsilon>0\exists\delta>0\forall y\in\underset{\sim}{R}(|y-x|<\delta\ \&\ y\#x\to\left|\frac{f(y)-f(x)}{y-x}-f'\right|<\varepsilon).$$

(B) f has a sequential derivative f' at x when

$$\forall x\forall\langle x_n\rangle_n(\forall n(x_n\#x)\ \&\ \lim_{n\to\infty}x_n=x\to\exists n\forall m\geq n(\left|\frac{f(x_m)-f(x)}{x_m-x}-f'\right|<\varepsilon)).$$

Then (A)↔(B).

Proof: (adapted from van Rootselaar 1952): (A)→(B) is immediate. To show (B)→(A) we construct a spread as follows. For each n, we define a sequence of rationals $\langle r^n_i\rangle_i$ such that

$$\begin{cases}2^{-n-1}\leq|r^n_i-x|\leq 2^{-n+1}\\ 2^{-n}<|r_j-x|<2^{-n}+2^{-n-1}\to\exists i(r_j=r^n_i)\end{cases}$$

($\langle r_n\rangle_n$ a standard enumeration of the rationals). Now we consider the collection of all sequences $\langle r^i_{\alpha i}\rangle^\infty_{i=0}$; all these sequences converge to x, but all elements are apart from 0. Note that the possible continuations of an initial segment $\langle r^0_{\alpha 0},\ldots,r^i_{\alpha i}\rangle$ do not depend on this initial segment.

Applying $\forall\alpha\exists x$-continuity to the statement

$$\forall\alpha\exists n\forall m\geq n(\left|\frac{f(r^m_{\alpha m})-fx}{r^m_{\alpha m}-x}-f'\right|<2^{-\nu})$$

(ν fixed), we find natural numbers k,p such that

$$\forall\beta\in\overline{\lambda y.0}(k)\forall m\geq p(\left|\frac{f(r^m_{\beta m})-fx}{r^m_{\beta m}-x}-f'\right|<2^{-\nu}).$$

Now we use the fact that, if we take $p=k$, then $r^m_{\beta m}, m\geq p$ ranges over a species containing all rationals $r, r\#x, |r-x|<2^{-k}$.

Therefore

$$\forall n(|r_n-x|\ <\ 2^{-k}\ \&\ r_n\#x\ \to\ \left|\frac{fr-fx}{r-x}-f'\right|\ <\ 2^{-\nu}).$$

Now consider any y, $|x-y|<2^{-k}$, $x\#y$. y is the limit of a sequence $\langle r_{\alpha n}\rangle_n$, $|x-r_{\alpha n}|<2^{-k-1}$ for all n, $x\#r_{\alpha n}$ for all n. Hence

$$\exists m\forall n\geq m\left(\left|\frac{fr_{\alpha n}-fx}{r_{\alpha n}-x}-f'\right| < 2^{-\nu}\right).$$

Also f is continuous on $\underset{\sim}{R}$ (application I), and therefore $\frac{fz-fx}{z-x}$ is continuous as a function of z at $z=y$, $y\#x$. As a consequence,

$$\forall n\geq m'\left(\left|\left(\frac{fr_{\alpha n}-fx}{r_{\alpha n}-x}\right)-\left(\frac{fy-fx}{y-x}\right)\right| < 2^{-\nu}\right)$$

for suitable m', therefore

$$\left|\frac{fy-fx}{y-x}-f'\right| < 2.2^{-\nu}$$

for all $y\#x$, $|y-x|<2^{-k}$.

6.6. Application V: Covering theorem.

Let $\Gamma=\langle V,\rho\rangle$ be a complete, separable, metric space, and let $\{V_i:i\in I\}$, $I\subset N$ be a numerically indexed covering of Γ. Then $\{Int\,V_i:i\in I\}$ also covers Γ ($\text{Int}W \equiv_{\text{def}}$ Interior of W).

Proof: Troelstra (1969), Thm. 6. A complete, separable, metric space Γ can be represented as follows. Let $\langle p_n\rangle_n$ be a basis for Γ, ρ its metric. Let $\lambda m.b(k,n,m)$ enumerate $\{i:\rho(p_n,p_i)<2^{-k}\}$; we may assume

$$p_n=p_m \rightarrow \forall kl(b(k,n,l)=b(k,m,l)).$$

The enumerating function b can be constructed using the function a approximating the metric ρ:

$$|\rho(p_n,p_m)-r_{a(n,m,k)}| < 2^{-k} ;$$

we only have to enumerate

$$\{m:\exists l(r_{a(n,m,l)}<2^{-k}-2^{-l})\}$$

for if $\rho(p_n,p_m)<2^{-k}$, then there is an l such that $\rho(p_n,p_m)<2^{-k}-2^{-l+1}$ and also $|\rho(p_n,p_m)-r_{a(n,m,l)}|<2^{-l}$, hence $r_{a(n,m,l)}<2^{-k}-2^{-l+1}+2^{-l}=2^{-k}-2^{-l}$; and conversely if $r_{a(n,m,l)}<2^{-k}-2^{-l}$, $\rho(p_n,p_m)<2^{-k}$.

Now we let correspond to a sequence α the sequence

$\langle q_n\rangle_n$ of basis points such that if $\delta 0 = \alpha 0$, $\delta(n+1) = b(n,\delta n,\alpha n)$, then $\forall n(q_n = p_{\delta n})$. Let us write x_α for $\lim\langle q_n\rangle_n$.

We then verify that each $x \in V$ coincides with an x_α, and that in fact (writing $U(\varepsilon,x)$ for the ε-neighbourhood of x) for each x there is an α such that

(1) $\quad x_\alpha = x \;\&\; \forall k(U(2^{-k-2},x) \subset W_{\overline{\alpha}k})$

where $W_n = \{x_\beta : \beta \in n\}$. To see this, consider any x, and let $\langle q_n\rangle_n \subset \langle p_n\rangle_n$ be such that $\rho(q_n,x) < 2^{-n-2}$ for all n, hence $\forall n(\rho(q_n,q_{n+1}) < 2^{-n-1})$. Then for some α, with $\delta 0 = \alpha 0$, $\delta(n+1) = b(n,\delta n,\alpha n)$, we have $\forall n(q_n = p_{\delta n})$, so $x = x_\alpha$. Let $\rho(x,y) < 2^{-k-2}$, and let $\langle q'_n\rangle_n \subset \langle p_n\rangle_n$ be such that $\forall n(\rho(q'_n,y) < 2^{-n-2})$, then $\forall n(\rho(q'_n,q'_{n+1}) < 2^{-n})$, $\rho(q'_k,q_{k-1}) \le \rho(q'_k,y) + \rho(y,x) + \rho(x,q_{k-1}) < 2^{-k-2} + 2^{-k-2} + 2^{-k-1} = 2^{-k}$, hence $q_0,q_1,\ldots,q_{k-1},q'_k,q'_{k+1},\ldots$ has a limit x_β with $\overline{\beta}k = \overline{\alpha}k$; $x_\beta = y$, hence $y \in W_{\overline{\alpha}k}$.

Now apply the weak continuity schema

WC-N $\quad \forall\alpha\exists x A(\alpha,x) \to \forall\alpha\exists x\exists y\forall\beta\in\overline{\alpha}y\, A(\beta,x)$

to the statement

$$A(\alpha,n) \equiv (n \in I \;\&\; x_\alpha \in V_i);$$

then we find

(2) $\quad \forall\alpha\exists n\exists k(W_{\overline{\alpha}k} \subset V_n \;\&\; n \in I)$.

Now with (1) we see that $\forall x\exists n{\in}I(x \in \mathrm{Int}(V_n))$.

Note that the index set I may depend on choice parameters.

An interesting consequence of this theorem is the following

<u>Corollary.</u> The apartness relation # on a complete separable metric space $\Gamma = \langle V,\rho\rangle$ given by $x\#y \equiv_{\mathrm{def}} \rho(x,y) > 0$ (which implies, intuitionistically, $\exists k(\rho(x,y) > 2^{-k})$) is explicitly definable in terms of equality by $x\#y \leftrightarrow \forall z(x \neq z \vee y \neq z)$.

Proof: Let $V_x = \{z : z \neq x\}$, $V_y = \{z : z \neq y\}$; assume $\forall z(x \neq z \vee y \neq z)$. Then $V \subset V_x \cup V_y$, and so $V \subset \mathrm{Int}(V_y) \cup \mathrm{Int}(V_y)$. Obviously $x \notin V_x$, hence $x \in \mathrm{Int}(V_y)$; similarly $y \in \mathrm{Int}(V_x)$. Therefore there is a k such that $U(2^{-k},x) \subset V_y$, and thus $\rho(x,y) > 2^{-k-1}$.

6.7. Application VI: Lindelöf's covering theorem.

Let $\Gamma = \langle V,\rho\rangle$ be a separable, complete metric space, and let $\{V_i : i \in I\}$ be a covering by open sets. Then there is a countable subcovering of $\{V_i : i \in I\}$.

Proof: We now have to use the continuity schema in the slightly stronger form

$$\forall\alpha\exists x A(\alpha,x) \rightarrow \exists\beta\in K_0 \forall\alpha A(\alpha,\beta(\alpha))$$

applied to the statement

$$\forall\alpha\exists k\exists i\in I(x_\alpha \in U(2^{-j_1 k}, p_{j_2 k}) \subset V_i).$$

Here x_α is the point of Γ represented by α as in the preceding application, $\langle p_n\rangle_n$ is the basis of Γ, $U(\varepsilon,x)$ indicates the ε-neighbourhood of x. From the statement we find an enumerating function γ (obtained via the neighbourhood function β in the continuity schema which gives us k in terms of α) such that

$$\forall\alpha\exists n\exists i\in I(x_\alpha \in U(2^{-j_1 \gamma n}, p_{j_2 \gamma n}) \subset V_i)$$

and also

$$\forall n\exists i\in I(U(2^{-j_1 \gamma n}, p_{j_2 \gamma n}) \subset V_i).$$

Applying a selection principle yields a $\psi \in N \rightarrow I$ such that

$$\forall\alpha\exists n(x_\alpha \in V_{\psi n})$$

and $\{V_{\psi n} : n \in N\}$ is the required countable subcovering.

6.8. Application VII: the Heine-Borel covering theorem.

This application is also 'classic' (cf. e.g. Brouwer (1926) or Heyting (1956)), and may be regarded as a specialization of the preceding two applications to compact spaces, now yielding finite coverings.

For closed intervals of $\mathbb{R}$, the theorem may be stated as follows:

Theorem. Let $\{V_i : i \in I\}$ cover $[a,b]$, and either V_i open for all i, or $I \subset N$. Then there is a finitely indexed subcovering $V_{i_1},\ldots,V_{i_n}$ such that $[a,b] \subset \mathrm{Int}(V_{i_1}) \cup \ldots \cup \mathrm{Int}(V_{i_n})$. For the proof, see e.g. Heyting (1956) 5.2.2.

Not only is it possible to recover Application I from

this theorem but it has many consequences, such as the following

Theorem. (Brouwer, Collected Works, page 558). A sequence of functions $\langle f_n \rangle_n$ defined on $[0,1]$ which converges for every $x \in [0,1]$ converges uniformly to a limit f on $[0,1]$.

Proof: Let $f(x) = \lim_{n\to\infty} f_n(x)$. Not only is $f(x)$ itself uniformly continuous, but since the set of W_n:

$$W_n = \{x : \forall m(|f_{n+m}(x) - f(x)| < 2^{-k})\}$$

covers $[0,1]$, there is a finite subset $W_{n_1},\ldots,W_{n_p}$ which already covers $[0,1]$; hence for $n = \max(n_1,\ldots,n_p)$ W_n covers $[0,1]$, i.e. $\forall x \in [0,1] \forall m(|f_{n+m}(x) - f(x)| < 2^{-k})$

6.9. Application VIII: Equivalence of two definitions of strong inclusion.

We define for metric spaces

$$V \subset\subset W \equiv_{\text{def}} \forall x \exists k(U(2^{-k},x) \cap V = \emptyset \vee U(2^{-k},x) \subset W)$$

and

$$V \text{ is located } \equiv_{\text{def}} \forall x \forall k(\exists y(y \in U(2^{-k},x) \cap V) \vee \exists l > k(U(2^{-l},x) \cap V = \emptyset));$$

V^- indicates the closure of V.

Then we have the theorem

If V,W are pointsets of a metric space, V located, then $V \subset\subset W \leftrightarrow V^- \subset \text{Int}(W)$.

Proof: (Troelstra 1968A): If $V \subset\subset W$, assume $x \in V^-$; then $\forall k \exists y(y \in V \cap U(2^{-k},x))$, and therefore for some k (using $V \subset\subset W$) $U(2^{-k},x) \subset W$, hence $x \in \text{Int}(W)$. Conversely, let $V^- \subset \text{Int}(W)$, V located, x arbitrary.

Let $\langle x_n \rangle_n$ be a sequence of points and let α be a sequence such that

$$\forall n((x_n \in U(2^{-\alpha n},x) \cap V) \vee (U(2^{-\alpha n},x) \cap V = \emptyset)),$$

$$\forall n \forall m(n > m \rightarrow \alpha n > \alpha m).$$

If $U(2^{-\alpha 0},x) \cap V = \emptyset$, we have finished. So assume $x_0 \in U(2^{-\alpha 0},x) \cap V$. We construct a tree of functions T according to the following picture:

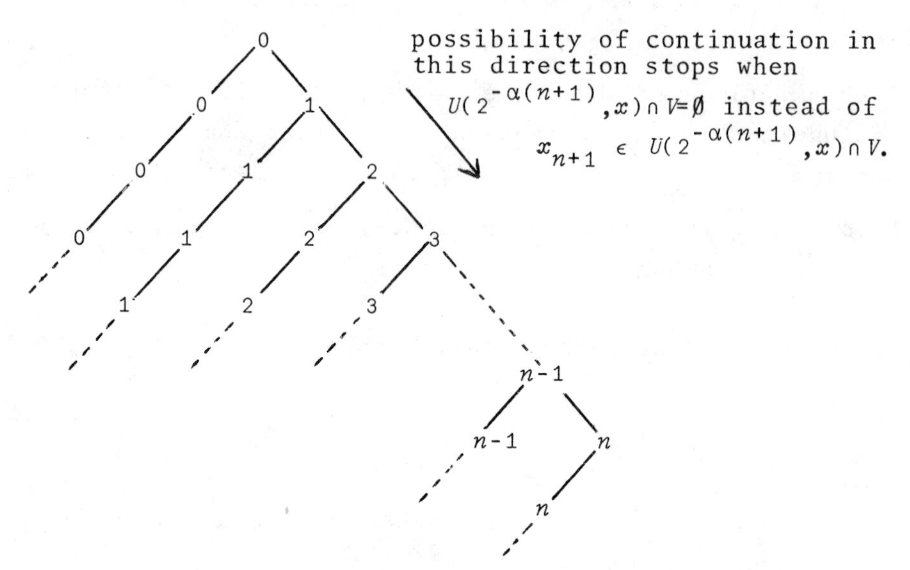

More precisely: all non-empty initial segments of sequences belonging to the tree are of the form $(\overline{\lambda x.x})(y+1)*(\overline{\lambda x.y})z$; and such a sequence is permitted only when $x_y\in U(2^{-\alpha y},x)\cap V$.

We can easily find a continuous functional represented by a neighbourhood function f such that $\forall\alpha\in T(f|\alpha=\alpha),\forall\alpha(f|\alpha\in T)$. Then we obtain weak continuity for T:

$$\forall\alpha\in T\exists xA(\alpha,x)\to\forall\alpha\in T\exists x\exists y\forall\beta\in T(\overline{\beta}x=\overline{\alpha}x\to A(\beta,y))$$

from ordinary weak continuity, since $\forall\alpha\in T\exists xA(\alpha,x)$ implies $\forall\alpha\exists xA(f|\alpha,x)$, hence by ordinary weak continuity $\forall\alpha\exists x\exists y\forall\beta(\overline{\beta}x=\overline{\alpha}x\to A(f|\beta,y))$, which on restriction of α,β to T yields the desired conclusion. We note that if $\alpha\in T$, then $\forall n(x_{\alpha n}\in V)$, and $\langle x_{\alpha n}\rangle_n$ converges, i.e. its limit belongs to V^-.

Now we may in fact assume, without restriction, that $V=V^-$, since V located $\leftrightarrow V^-$ located, and $V\subset\subset W\leftrightarrow V^-\subset\subset W$, so then $\lim_n x_{\alpha n}\in V$. Now from $V=V^-\subset\mathrm{Int}(W)$ we conclude

$$\forall\beta\in T\exists k(U(2^{-k},\lim_n x_{\beta n})\subset W);$$

applying weak continuity for T to this statement yields

$$\forall\gamma\in T\exists m\exists k\forall\beta\in T(\overline{\gamma}m=\overline{\beta}m\to U(2^{-k},\lim_n x_{\beta n})\subset W).$$

We apply this conclusion to γ=rightmost branch of the spread, and find m',k' such that

$$(1)\qquad \forall\beta\in T(\overline{\gamma}m'=\overline{\beta}m'\to U(2^{-k'},\lim_n x_{\beta n})\subset W).$$

Now take $k=\max(k',m')$, then also

$$\forall\beta\in T(\overline{\gamma}(k+1)=\overline{\beta}(k+1) \rightarrow U(2^{-k}, \lim_n x_{\beta n})\subset W).$$

Take for β: $\gamma 0,\gamma 1,\ldots,\gamma(k-1),\gamma k,\gamma k,\gamma k\ \gamma k,\ldots$. There are two possibilities: either $U(2^{-\alpha i},x)\cap V=\emptyset$ for some $i\leq k$, or $x_k\in U(2^{-\alpha k},x)\cap V$. In the first case $\exists k(U(2^{-k},x)\cap V=\emptyset)$. In the second case we note that also $U(2^{-k},x_k)\subset W$ (since $x_k=\lim_n x_{\beta n}$ in this case, and (1) holds); and since we can find an l such that $\rho(x,x_k)<2^{-k}-2^{-l}$, it follows that $U(2^{-l},x)\subset U(2^{-k},x_k)\subset W$. Hence in this case $\exists l(U(2^{-l},x)\subset W)$, and therefore always

$$\exists\varepsilon(U(\varepsilon,x)\subset W \vee U(\varepsilon,X)\cap V=\emptyset)$$

i.e. $V\subset\subset W$.

6.10. *Application IX: 'Sharp difference' implies apartness.*

If x,y,z are real numbers, we define

$$x<y \equiv_{\text{def}} \exists k(y-x>2^{-k})$$

$$z\in[x,y] \equiv_{\text{def}} \min(x,y) \nmid z \nmid \max(x,y)$$

$$z\in(x,y) \equiv_{\text{def}} x\neq z \ \&\ y\neq z \ \&\ z\in[x,y]$$

$$x\neq_s y \equiv_{\text{def}} x\neq y \ \&\ \forall z\notin(x,y)[(z\nmid x \ \&\ z\nmid y) \vee (z\nless x \ \&\ z\nless y)]$$

(x is sharply different from y).

For this somewhat artificial notion we have

$$x\neq_s y \rightarrow x\# y,$$

(Kleene and Vesley (1965) page 168), a theorem of historical interest only (first stated in Brouwer 1930).

6.11. *Application X: Riemann's permutation theorem.*

The following theorem is well known from classical analysis:

(a) If the series $\Sigma_{n=0}^{\infty}x_n$ is absolutely convergent, then $\Sigma_{n=0}^{\infty}x_n=\Sigma_{n=0}^{\infty}x_{\sigma(n)}$ for each permutation σ of the natural numbers (a permutation of N is a bi-unique mapping of N onto N).

(b) If $\Sigma_{n=0}^{\infty}|x_n|$ is divergent, but $\Sigma_{n=0}^{\infty}x_n$ convergent, then for each real number x there exists a permutation σ such that $\Sigma_{n=0}^{\infty}x_{\sigma(n)}=x$; there also exist permutations σ',σ'' such that $\Sigma_{n=0}^{\infty}x_{\sigma'(n)}=+\infty$, $\Sigma_{n=0}^{\infty}x_{\sigma''(n)}=-\infty$.

Classically, we immediately get from (b):

(c) If $\Sigma_{n=0}^{\infty} x_{\sigma(n)}$ is convergent for all permutations σ, then $\Sigma_{n=0}^{\infty} x_n$ is absolutely convergent.

The classical proof simply consists in applying contraposition to statement (b). Statement (b) can also be proved intuitionistically, provided divergence is interpreted intuitionistically as $\forall k \exists m (\Sigma_{n=0}^{m} |x_n| > k)$; but this is not enough to get an intuitionistic proof of (c), and therefore we need a separate intuitionistic proof of (c). As regards (a) the standard classical proof is also valid intuitionistically.

Proof[†] *of* (c): The reader who only desires to see how the continuity schema is applied can restrict himself to the part of the proof where (c) is established under simplifying assumptions, and omit the part where the assumptions are eliminated.

We have to bring in a suitable $\forall\alpha\exists x$-statement to apply our continuity schemas to; and the obvious idea is to let the permutations of the natural numbers be coded by a tree of sequences of natural numbers. The simple-minded idea of letting the finite sequences of the tree simply consist of the initial segments of permutations does not work; although it is decidable whether a finite sequence can occur as initial segment of a permutation (it is sufficient that all elements of the sequence are distinct), a sequence with all its initial segments possible initial segments of a permutation function need not be itself a permutation: 0,2,4,6, 8,10,... is a counterexample. The root of the trouble is that we must be assured, not only that $\sigma(x)$ is determined for all x, but also that each x occurs in the image of N under σ, i.e. $\sigma^{-1}(x)$ must be determined for each x. To obtain a suitable coding by a tree of sequences, we use a

---

[†] The result to be proved was announced in the abstract of Gielen (1969), without proof. We have not seen the proof, but presumably it is along the same lines as the one given here.

sequence m of length n to code $\sigma(0),\ldots,\sigma(n-1)$ and $\sigma^{-1}(0),\ldots,\sigma^{-1}(n-1)$, by taking care that $j(\sigma(i),\sigma^{-1}(i))=(m)_i$ for $i<n$. Now it is already fairly obvious that the condition for a finite sequence to code a piece of a permutation is decidable, but let us describe the condition explicitly:

$\langle\rangle$ is admissible;

$\langle x\rangle$ is admissible iff $(j_1x=0\leftrightarrow j_2x=0)$.

Assume $n=\langle x_0,\ldots,x_u\rangle$ to be admissible, then

$n*\hat{x}$ is admissible iff

$$\begin{cases} u+1=j_2x_i \quad \text{for} \quad i\leq u \rightarrow j_1x=i \\ u+1=j_1x_i \quad \text{for} \quad i\leq u \rightarrow j_2x=i \\ \text{and } j_1x=u+1 \leftrightarrow j_2x=u+1. \end{cases}$$

For any σ, let α_σ be the representing function, i.e. $\alpha_\sigma=\lambda x.j(\sigma x,\sigma^{-1}x)$. To the tree of α_σ we may apply weak continuity (cf. application VII).

Proof (under simplifying assumptions):
We first show how to obtain the theorem under the simplifying assumptions

(1) $\forall i(x_i>0 \vee x_i<0 \vee x_i=0)$,

(2) $\exists x(\Sigma_{n=0}^{\infty}x_{\sigma n}=x)$ for all σ.

Let $x=\Sigma_{n=0}^{\infty}x_n$. Note that $\alpha_\sigma=\lambda x.j(x,x)$ in the case that σ is the identity mapping. By assumption (2) it follows that

$$\forall\alpha_\sigma\exists m\forall u\geq m(|(\Sigma_{n=0}^{u}x_{\sigma n})-x|<2^{-\nu})$$

for each fixed ν. Applying weak continuity to this statement we obtain:

(3) $\forall\alpha_\sigma\exists m\exists k\forall\alpha_\tau(\overline{\alpha}_\sigma(k)=\overline{\alpha}_\tau(k) \rightarrow \forall u\geq m(|(\Sigma_{n=0}^{u}x_{\tau n})-x|<2^{-\nu}))$.

Choose $\lambda x.j(x,x)$ for α_σ, and take $m+1$ and k to be equal (this is always possible), then for some k

$$\forall\alpha_\tau(\overline{\alpha}_\tau(k+1)=\overline{\lambda x.j(x,x)}(k+1) \rightarrow \forall u\geq k(|(\Sigma_{n=0}^{u}x_{\tau(n)})-x|<2^{-\nu})).$$

Let $t(0),\ldots,t(l)$ be the indices i of the first $l+1$ terms x_i such that $x_i>0$ and $i>k$, with $k<t(0)<t(1)<\ldots<t(l)$. Then we can find a permutation τ with

$\overline{\alpha}_\tau(k+1)=\overline{\lambda x.j(x,x)}(k+1)$

$\tau(k+i+1)=t(i)$ for $0\leq i\leq l$.

For this τ it follows that, choosing $k+l+1,k$ for u,k,

$$|\Sigma_{n=0}^{k+l+1} x_{\tau n}-x|<2^{-\nu}, |\Sigma_{n=0}^{k} x_{\tau n}-x|<2^{-\nu}$$

and therefore

$$|\Sigma_{n=k+1}^{k+l+1} x_{\tau n}|<2.2^{-\nu}.$$

This holds for arbitrary l, therefore the subseries of positive terms x_i converges. Similarly for the subseries of negative terms; therefore also $\Sigma_{n=0}^{\infty}|x_n|$ converges.

Removal of the simplifying assumptions.

We shall first eliminate assumption (2). (3) is to be replaced by

(4) $\quad \forall\alpha_\sigma\exists l\exists x\forall u\geq l(|(\Sigma_{n=0}^{u} x_{\sigma n})-x|<2^{-\nu}).$

In order to be able to apply the continuity schema, we must rewrite (4) as

(5) $\quad \forall\alpha_\sigma\exists l\exists i\exists x\forall u\geq l(|(\Sigma_{n=0}^{u} x_{\sigma n})-x|<2^{-\nu}\ \&\ |r_i-x|<2^{-\nu}),$

where $\langle r_i\rangle_i$ is a standard enumeration of the rationals. Arguing along the same lines as before, we find

(6) $\quad \forall\alpha_\sigma\exists l\exists i\exists k\forall\alpha_\tau(\overline{\alpha}_\sigma k=\overline{\alpha}_\tau k \to \forall u\geq l(|(\Sigma_{n=0}^{u} x_{\tau n})-r_i|<2.2^{-\nu})),$

and then copying the argument given above

$$|\Sigma_{n=k+1}^{k+1+l} x_{\tau n}|<4.2^{-\nu}$$

($\alpha_\sigma,\alpha_\tau$ chosen as before) etc.

We can eliminate assumption (1) as follows. Each x_i is the limit of a sequence of rationals $\langle r_{i,j}\rangle_j$ with a prescribed rate of convergence

$$\forall ij(|x_i-r_{i,j}|<2^{-j}).$$

Now $\langle r_{i,i+1}\rangle_i$ is a sequence such that convergence of $\Sigma_{n=0}^{\infty}|x_n|$, $\Sigma_{n=0}^{\infty} x_{\sigma n}$ is equivalent to convergence of $\Sigma_{n=0}^{\infty}|r_{n,n+1}|$, $\Sigma_{n=0}^{\infty} r_{\sigma n,\sigma n+1}$ respectively, for each permutation σ. To see this, observe

$$\Sigma_{n=k}^{k+l}|x_n-r_{n,n+1}|<2^{-k}$$

and that for any k and any σ such that $0,1,\ldots,m$ occur among $\sigma 0,\ldots,\sigma(k-1)$ it follows that for each l

$$\Sigma_{n=k}^{k+l}|x_{\sigma n}-r_{\sigma n,\sigma n+1}|<2^{-m}.$$

Therefore if $\langle x_n\rangle_n$ satisfies the premiss of (c), so does

$\langle r_{n,n+1} \rangle_n$; moreover, $\langle r_{n,n+1} \rangle_n$ satisfies assumption (1).

6.12. The use of the elimination mappings for constructivizing theorems.

As we have seen in our examples above, various classical theorems relying for their classical proofs on classical logic (such as the implication sequential continuity ⇒ continuity, in application III, and Riemann's permutation theorem in application X) can be proved intuitionistically with the help of the continuity axioms. Suppose we interpret sequences in the intuitionistic version as choice sequences in the sense of $\underset{\sim}{\mathrm{CS}}$, it follows that the translation under σ (the elimination mapping of Chapter 5), is provable in $\underset{\sim}{\mathrm{IDB}}_1$, i.e. the translation is a theorem of constructive mathematics in the narrow sense of Bishop (1967) ('narrow' = 'not involving choice sequences'). Let us take Riemann's permutation theorem as an illustration of this method of constructivization.

It can be read off from the proof of the permutation theorem that the formula (6) for all ν implies absolute convergence without further continuity assumptions. With a bit of rewriting this yields the following:

Theorem. If $\forall k \forall \sigma \exists x \exists m \forall \tau \forall n \forall n' (\overline{\sigma}x = \overline{\tau}x \to |\Sigma_{i=m+n}^{m+n+n'} x_{\tau i}| < 2^{-k})$, then Σx_n is absolutely convergent. Here σ, τ range over permutations of N.

This is in itself not a very useful theorem. However, it certainly seems worthwhile to hunt for further and more interesting examples. We should add that another well-known method for obtaining constructivizations, the Dialectica interpretation (see Troelstra (1973) section 3.5 for a description) when applied to the permutation theorem leads to a rather complicated expression without a clear intuitive meaning. The advantage of the Dialectica interpretation in general is, of course, that it applies to proofs in classical logic, whereas the method indicated here applies to theorems provable with intuitionistic logic and continuity. But such a method of rather restricted applicability may yield better results in the cases where it applies, as is illustrated by Riemann's permutation theorem.

Interpreting the sequences as projections of lawless sequences in the sense of Chapter 4 automatically induces other mappings; for example, interpretation in $\mathcal{U}^*_\beta$ leads to replacing quantifiers $\forall\alpha$, $\exists\alpha$ by $\forall e$, $\exists e$, and the occurrences of α by $e|\beta$, then quantifying over β and applying the τ for lawless sequences. For $\mathcal{U}$ it is even more complicated. However, these translations are less promising than σ, precisely because assertions about the α are replaced by assertions about their components (e, β, for all β in the case of $\mathcal{U}^*_\beta$, and e, ε^u in the case of $\mathcal{U}$) whereas mathematically the α's (i.e. the $e|\beta$'s, $e|\varepsilon^u$'s respectively) are the ones which interest us. Note also that the properties expressible in terms of α only which can be used to prove e.g. the permutation theorem relative to $\mathcal{U}$, are simply valid in $\underset{\sim}{\mathrm{CS}}$ too (see the remarks in section 6.1).

7
CHOICE SEQUENCES AND COMPLETENESS OF INTUITIONISTIC PREDICATE LOGIC

7.1. *Introduction.*

The aim of this chapter is to discuss in some detail completeness for intuitionistic first order predicate logic IPC for the intuitive concept of validity (in the sense of being intuitionistically true for all intuitionistically meaningful domains and relations over these domains).

Conventions. In order to simplify the discussion, we restrict attention to *pure* first order predicate logic IPC, without equality and without function symbols; we leave open the possibility that the language contains constants. There is a countable list of predicate symbols $P_0, P_1, P_2, \ldots$; P_0 is identified with $\bot$. P_n has $\tau(n)$ arguments (τ a primitive recursive function of n). The logical constants are $\forall, \exists, \to, \&, \vee, \bot$ (absurdity); $\neg A$ is an abbreviation for $A \to \bot$. When a notational distinction is necessary, we use $\underset{\sim}{\exists}, \underset{\sim}{\forall}, \Rightarrow, \Leftrightarrow$, and, or, not in our metalanguage.

Definition. Let $A(P_1, \ldots, P_n)$ be a formula of IPC, with all its predicate letters among $P_1, \ldots, P_n$. Let D range over (intuitionistically meaningful) domains, P_i^* $(1 \le i \le n)$ over relations over D (where P_i^* and P_i have the same number of arguments), and let A^D be obtained from A by relativizing quantifiers to D, and substituting P_i^* for P_i. Then validity in all structures (or intuitive validity) is defined by

$$\mathrm{Val}(A) \equiv_{\mathrm{def}} \underset{\sim}{\forall} D \underset{\sim}{\forall} P_1^* \ldots \underset{\sim}{\forall} P_n^* A^D(P_1^*, \ldots, P_n^*).$$

Val_c is defined as Val, but the range of D, P_i^* is restricted to domains and relations not depending on non-lawlike parameters (e.g. choice sequences).

We define completeness (Comp) and weak completeness (Comp') with respect to A by ($\mathrm{Proof}_{\mathrm{IPC}}$ being the formalized proof predicate)

$$\mathrm{Comp}(A) \equiv_{\mathrm{def}} \mathrm{Val}(A) \to \exists x\, \mathrm{Proof}_{\mathrm{IPC}}(x, \ulcorner A \urcorner),$$

$$\mathrm{Comp}'(A) \equiv_{\mathrm{def}} \mathrm{Val}(A) \rightarrow \neg\neg\exists x\ \mathrm{Proof}_{\mathrm{IPC}}(x, \ulcorner A \urcorner).$$

Comp_C, Comp'_C are the corresponding notions with Val_C instead of Val.

Discussion of Val. The notion Val for IPC is often rejected on the grounds that it is 'too vague' - in particular, the extension of the class of (intuitionistically meaningful) domains and predicates is insufficiently known. However, a comparison with the case of CPC (= classical first order predicate logic) shows that the situation with respect to validity and completeness is basically similar in both cases. For CPC, the situation may be summed up in three points:

(I) $\exists x\ \mathrm{Proof}_{\mathrm{CPC}}(x, \ulcorner A \urcorner) \Rightarrow \mathrm{Val}(A)$. Here $\mathrm{Val}(A)$ is defined as for the intuitionistic case, now for classically meaningful domains and relations. The structures (i.e. domains with relations over them) to which $\mathrm{Val}(A)$ refers, are *not necessarily set-theoretic*. Here 'set-theoretic' means: represented by elements of the 'intended' model (e.g. the cumulative hierarchy of types) of ZF. This implication is *obvious*, whatever the precise extension of the concept of a structure may be.

(II) $\mathrm{Val}(A) \Rightarrow A$ is valid in all set-theoretical structures. This implication is again obvious. But the converse is not, since there is a natural example of a structure which is not set-theoretic: formulae of IPC with a single binary relation E admit an interpretation where the domain consists of the class of all sets (and hence is not a set) and E is set-theoretic membership. For a detailed discussion we refer to Kreisel (1967) section 2.

(III) For a converse to (I), i.e. completeness in the sense of Comp or Comp' (which are classically equivalent notions), we expect that any proof of such a converse yields, on inspection, more than is actually stated: namely that it suffices to restrict attention to a certain special class of structures which can be explicitly described. The proof of the completeness theorem for CPC shows that we can in fact restrict attention to Δ_2^0-definable structures (hence a fortiori to arithmetical or set-theoretic structures). Thus we see that the exact extension of the concept of structure is irrelevant, since a quite limited class of structures suffices.

Suppose the converse to (I) were *false*, then a proof of this fact would take the following form: Under certain quite specific assumptions (closure conditions) regarding the class of structures, incompleteness holds; in other words, for all classes of structures satisfying the closure conditions we would have had incompleteness. For example, the restriction to recursively enumerable structures (structures where domain and relations are Σ_1^0-definable), we have *incompleteness* of CPC (see e.g. Putnam (1957), Vaught (1960)).

For IPC, the situation is similar, but more complicated; for a summary of results, see Troelstra (1976) section 3. Obviously, the extension of the class of valid sentences depends on the mathematical assumptions about the possible structures. Classically, Val is determined by the extension of the class of domains and relations; intuitionistically, Val not only depends on this extension, but also on the class of proofs of logically compound statements concerning intuitionistic structures. This dependence on the class of proofs is implicit in the axioms we postulate for certain mathematical objects; for example, where relations with a lawless parameter are included in the range of the relation quantifiers in the definition of Val, the extension of Val is determined by the axioms for lawless sequences, and the axioms express something about the possible proofs for statements of a certain form (see for an illustration section 7.13). As a result, the dependence of Val on mathematical assumptions is much more striking and essential in the intuitionistic case. For example, on the more or less plausible assumption of Church's thesis, IPC is *incomplete* (see the exposition in van Dalen 1973).

By stressing the similarity between the classical and the intuitionistic situation, we do not wish to claim equal interest for Val in both cases; in the classical case, the study of Val gave rise to model theory; in the intuitionistic case, the most fruitful notion so far was that of a Kripke model, treated *classically*; and validity in Kripke models is only indirectly connected with Val. Another factor which limits the interest of the existing completeness results for Val is the fact that they require the consideration of relations with lawless parameters, i.e. the structures needed are unfamiliar and not directly related to mathematical practice - in contrast to the classical case. Nevertheless the study of Val is of interest, also in the intuitionistic case, because of the naturalness of Val, and as an illustration of the inseparability of logic and mathematics.

Description of the contents of the chapter. We first of all present an intuitionistic completeness proof relative to 'generalized Beth models'. Roughly speaking, in order to make the collection of Beth models into a neat topological space, we have to embed the Beth models in the proper sense in a larger class containing also possibly 'inconsistent' models.[†]

We pay special attention to the properties of the functions coding generalized Beth models which are needed in the proof. Then we extract from this proof the original result of Dyson and Kreisel (1961), and show the connection between validity in Beth models and validity in the sense of being intuitionistically true in domains and relations over these domains containing lawless parameters. The final result is summarized in section 7.13 (in the theorem and its corollaries). The use of Beth models and a Gödel-style completeness proof might have been replaced by the use of

---

[†]This idea originated with E. W. Beth (see the historical note, section 7.17). We gratefully acknowledge H. de Swart's permission to use his unpublished report (de Swart 1974) for our presentation.

Kripke models with a Henkin-style completeness proof (cf. Lopez-Escobar and Veldman 1975). Our choice was motivated by didactic reasons: the construction of the models is very easily visualized and readily lends itself to an axiomatic analysis of the properties of sequences needed in the proof, and the connection between Beth models (as special cases of topological models) and validity in terms of domains and relations is very direct. As a byproduct we get (for free) an intuitionistic model-theoretic proof of closure under cut. For a survey of results on validity in all structures, see Troelstra (1976).

7.2. Definition of generalized Beth model and Beth model.

Let $\langle T,\leq\rangle$ be a decidable finitely branching tree without finite branches, with partial ordering $\leq$. We shall identify $\langle T,\leq\rangle$ with a tree of finite sequences of natural numbers, with the obvious partial ordering. Then $\langle\rangle$ is the root of the tree, and $\forall n(n\in T \vee \neg n\in T)$, $\forall m\in T\forall n(n \leq m \rightarrow n\in T)$. We shall use $k,k',k'',\ldots$ as variables for elements of T.

A *generalized Beth model* is a triple $\langle T,\leq,\phi\rangle$, $\langle T,\leq\rangle$ a tree as above, and ϕ, the *model function*, is a mapping

$$\phi\colon T\rightarrow\mathfrak{P}\left(\bigcup_{i=0}^{\infty}\{i\} \times D^{\tau(i)}\right).$$

Here $\mathfrak{P}X$ denotes the power set of X (the set of all subsets of X). D is called the *domain* of the model.

If for some k, $\langle 0,\emptyset\rangle \in \phi(k)$, the model is said to be *inconsistent*. A generalized Beth model that is not inconsistent is called a (proper) *Beth model*.

Remarks: (i) The definition may be considerably generalized by permitting infinitely branching trees, or even more liberally, arbitrary partially ordered systems as underlying partially ordered structure; then maximal chains take the place of infinite branches.
(ii) In the completeness proof we may restrict attention to the special case $D = \mathbb{N}$, and $\phi(k)$ *decidable* uniformly in k.
(iii) An inessential (but sometimes convenient) variant of the definition imposes the additional condition

$$k \leq k' \rightarrow \phi(k) \subset \phi(k').$$

7.3. Definition (of forcing).

We define, for any given generalized Beth model, with domain D, a relation $k \Vdash A$ ('k forces A' or 'the model forces A at k') for all $k \in T$ and sentences A of the language of $\underset{\sim}{\mathrm{IPC}}$ with constants for all elements of D. We use $d, d', \ldots$ as metavariables for such constants, and $\varepsilon, \varepsilon', \ldots, \eta, \eta', \ldots$ as variables for branches of T (we think of these branches as lawless sequences restricted to a fan; cf. section 2.18). The definition[†] is by induction on the logical complexity of A:

(i) $k \Vdash P_i(d_1, \ldots, d_{\tau(i)}) \equiv_{\mathrm{def}} \underset{\sim}{\forall} \varepsilon \in k \underset{\sim}{\exists} x(\langle i, d_1, \ldots, d_{\tau(i)} \rangle \in \phi(\bar{\varepsilon} x)$ <u>or</u> the model is inconsistent)

(ii) $k \Vdash A \& B \equiv_{\mathrm{def}} k \Vdash A$ <u>and</u> $k \Vdash B$

(iii) $k \Vdash A \vee B \equiv_{\mathrm{def}} \underset{\sim}{\forall} \varepsilon \in k \underset{\sim}{\exists} x(\bar{\varepsilon} x \Vdash A$ <u>or</u> $\bar{\varepsilon} x \Vdash B)$

(iv) $k \Vdash A \rightarrow B \equiv_{\mathrm{def}} \underset{\sim}{\forall} k' \geq k(k' \Vdash A \Rightarrow k' \Vdash B)$

(v) $k \Vdash \exists x A x \equiv_{\mathrm{def}} \underset{\sim}{\forall} \varepsilon \in k \underset{\sim}{\exists} x \underset{\sim}{\exists} d(\bar{\varepsilon} x \Vdash A d)$

(vi) $k \Vdash \forall x A x \equiv_{\mathrm{def}} \underset{\sim}{\forall} d(k \Vdash A d)$.

A sentence A is valid in the model $\langle T, \preceq, \phi \rangle$ if it is forced at $\langle\rangle$, i.e. $\langle\rangle \Vdash A$. Note that for all models

$$k' \geq k \text{ and } k \Vdash A \Rightarrow k' \Vdash A,$$

and for inconsistent models $k \Vdash A$ for all A, k. Both assertions follow by induction on the logical complexity of A, as does the following

<u>Lemma.</u> For all A of $\underset{\sim}{\mathrm{IPC}}$ and all generalized Beth models

$$k \Vdash A \Leftrightarrow \underset{\sim}{\forall} \varepsilon \in k \underset{\sim}{\exists} x(\bar{\varepsilon} x \Vdash A).$$

7.4. A formal system for $\underset{\sim}{\mathrm{IPC}}$ (without constants).

If A is a sentence, TA ('true-A') and FA ('false-A') are *signed* formulae. Let us use $S, S', S'', \ldots$ for finite sets

---

[†] Clause (i) is a slight modification, due to M. A. E. Dummett, of the original definition in de Swart (1974); with this definition the lemma below also applies to infinitely branching trees, which was not the case for the original definition.

of signed formulas. Furthermore we put, for any S:

$$S_T = \{TA : TA \in S\}$$

We now describe a (cut-free) calculus of sequents for IPC. A set of signed formulas $S = \{TA_1,\ldots,TA_n, FB_1,\ldots,FB_m\}$ is said to be a *sequent* (corresponding in the usual formulation of sequent calculi to $A_1,\ldots,A_n \vdash B_1,\ldots,B_n$; the intuitive interpretation is $(A_1\&\ldots\&A_n \to B_1\vee\ldots\vee B_m)$). Let us distinguish between (bound) variables v_0, v_1, v_2,... and parameters (free variables) a_0, a_1, a_2,... . We use x,y,z, u,v,w, and a,b,b_i as meta-variables for variables and parameters respectively.

Axioms: sequents S such that for some A TA, $FA \in S$ or $T\bot \in S$.

Rules :

$T\&$ $\dfrac{S,T(A\&B),TA,TB}{S,T(A\&B)}$ $F\&$ $\dfrac{S,FA \quad S,FB}{S,F(A\&B)}$

$T\vee$ $\dfrac{S,T(A\vee B),TA \quad S,T(A\vee B),TB}{S,T(A\vee B)}$ $F\vee$ $\dfrac{S,FA,FB}{S,F(A\vee B)}$

$T\to$ $\dfrac{S,T(A\to B),FA \quad S,T(A\to B),TB}{S,T(A\to B)}$ $F\to$ $\dfrac{S_T,TA,FB}{S,F(A\to B)}$

$T\exists$ $\dfrac{S,T(\exists xAx) \quad T(Aa)}{S,T(\exists xAx)}$ (a not in S) $F\exists$ $\dfrac{S,F(Aa)}{S,F(\exists xAx)}$

$T\forall$ $\dfrac{S,T(\forall xAx),T(Aa)}{S,T(\forall xAx)}$ $F\forall$ $\dfrac{S_T,F(Aa)}{S,F(\forall xAx)}$ (a not in S).

A proof tree starts from axioms and proceeds downwards by means of applications of the rules; the bottom sequent is said to be the conclusion of the proof tree. A is said to be derivable in IPC if there is a proof tree with $\{FA\}$ as conclusion.

7.5.

Theorem. (Soundness theorem). Let us define A to be valid in a Beth model $\langle K,\preceq,\phi\rangle$ iff $\langle\rangle \Vdash A$; *A is valid* if A is valid in all Beth models. A sequent $S = \{TA_1,\ldots,TA_n,FB_1,\ldots,FB_m\}$ is valid if for every Beth model in which $A_1,\ldots,A_n$ are valid, $B_1\vee\ldots\vee B_m$ is valid.
The soundness theorem states:

S derivable in $\underset{\sim}{IPC} \Rightarrow S$ valid.

Proof: Straightforward by induction on the length of deductions in $\underset{\sim}{IPC}$.

7.6. *Remarks on the formal system.*

Let us say we apply an *inverted rule* to a set of signed formulae S, if we replace S by one of the possible premisses S',S'' according to a rule of the system (i.e. $\frac{S',S''}{S}$ would be an application of one of the rules).

(i) Note that an inverted rule application can only *increase* the subset of T-formulae;

(ii) all formulae in the premise of a rule are subformulae of the formulae in the conclusion;

(iii) the non-active formulae (in $T\&$, $F\&$, $T\vee$,... the 'active' formulae are $T(A\&B)$, $F(A\&B)$, $T(A\vee B)$,... respectively) are always retained under application of an inverted rule *except* with inverted $F\rightarrow$, $F\forall$; there F-formulae may get lost.

Suppose we are searching systematically for derivations of a sequent S, by applying inverted rules to see whether we end up with a set of axioms. As long as we do not apply inverted $T\exists$, $F\forall$ or $F\rightarrow$, no possibilities get lost and there is no need to consider new parameters. Put differently, and more precisely, as long as we do not use inverted $T\exists$, $F\forall$, $F\rightarrow$ in our search, we do not have to introduce new parameters, and the order in which the inverted rules are applied is irrelevant - ultimately we end up with a collection of sets of signed formulae (just by applying inverted rules except inverted

$T\exists$, $F\forall$, $F\rightarrow$ as often as possible) which is independent of the order of application of the inverted rules; if all sets of this collection are *axioms*, we have actually found a proof tree (not using $T\exists$, $F\forall$, $F\rightarrow$).
But an application of e.g. $F\rightarrow$ *destroys* possibilities; e.g. if we consider

$$S,\ F(A\rightarrow B),\ F(C\rightarrow D)$$

there are two possibilities for applying $F\rightarrow$ inversely, yielding

$$S_T,\ TA,\ FB \quad \text{and} \quad S_T,\ TC,\ FD$$

respectively; but in the first case we have lost the possibility of treating $F(C\rightarrow B)$ and in the second case the possibility of treating $F(A\rightarrow B)$. We have to pursue the possibilities of finding a proof tree for S_T, TA, FB and S_T, TC, FD; it is *sufficient* to find a proof tree in *one* of these cases.

After these preliminary remarks, we are ready to describe more systematically the

7.7. Search procedure.

We say that a set Δ is a *Hintikka set* with respect to a set of parameters V, if

(i) $T(A\&B)\in\Delta \Rightarrow TA\in\Delta$ <u>and</u> $TB\in\Delta$

(ii) $T(A\vee B)\in\Delta \Rightarrow TA\in\Delta$ <u>or</u> $TB\in\Delta$

(iii) $T(A\rightarrow B)\in\Delta \Rightarrow FA\in\Delta$ <u>or</u> $TB\in\Delta$

(iv) $F(A\&B)\in\Delta \Rightarrow FA\in\Delta$ <u>or</u> $FB\in\Delta$

(v) $F(A\vee B)\in\Delta \Rightarrow FA\in\Delta$ <u>and</u> $FB\in\Delta$

(vi) $T(\forall xAx)\in\Delta \Rightarrow \underset{\sim}{\forall} a\in V(T(Aa)\in\Delta)$

(vii) $F(\exists xAx)\in\Delta \Rightarrow \underset{\sim}{\forall} a\in V(F(Aa)\in\Delta)$.

We call a Hintikka set Δ *minimal* over a set S, if all elements of S either belong to S or are present in virtue of (i)-(vii). Note that for any *finite* set S of signed formulae and *finite* set V, there are finitely many minimal Hintikka sets with respect to V containing S, each consisting of finitely many signed formulae. If all the Hintikka sets thus constructed from S are axioms, then there is a proof tree for S not using $T\exists$, $F\forall$ or $F\rightarrow$.

It is left to the reader to show: if all Hintikka sets

with respect to the parameters of S minimal over S are closed (i.e. contain $T\bot$ or TA, FA for some A), then there is a derivation of S in $\underset{\sim}{\text{IPC}}$ not using $F\!\to$, $F\forall$, $T\exists$. (Hint: note that any derivation in the calculus of sequents can be transformed into a derivation using only axioms containing $T\bot$ or TA, FA for *atomic* A.)

Let us call a Hintikka set Δ *full* if it also satisfies

(viii) $FA,\ TA \in \Delta \Rightarrow T\bot \in \Delta$

Now our search procedure consists of forming full Hintikka sets; pursuing for each such set the various possibilities for applying $F\!\to$, $T\exists$, $F\forall$ (of which only one needs to lead to a proof tree) again forming full Hintikka sets, etc. etc. More systematically, we now describe the process as follows. Let S be a finite set of signed formulae, and let Par(S) denote the collection of parameters in S.

Step 0. Construct all full Hintikka sets $\Gamma^{\langle i\rangle}_{\langle 0\rangle}$ with respect to Par(S), minimal over S.

Step 1. Each $\Gamma^{\langle i\rangle}_{\langle 0\rangle}$ may contain several *split* formulas (i.e. formulas of the form $F(A\to B)$ or $F(\forall x Ax)$), so there are various (disjunctive) possibilities.

Step 1a. To each $\Gamma^{\langle i\rangle}_{\langle 0\rangle}$ we construct a number of successors $\Gamma^{\langle i,j\rangle}_{\langle 0,m\rangle}$; the $\{\Gamma^{\langle i,j\rangle}_{\langle 0,m\rangle}\}_j$ minimal full Hintikka sets corresponding to the possibility of treating the mth split formula in $\Gamma^{\langle i\rangle}_{\langle 0\rangle}$.

(I) Let the mth split formula be of the form $F(B\to C)$, and assume $(\Gamma^{\langle i\rangle}_{\langle 0\rangle})_T \cup \{TB, FC\}$ to contain k formulas $T(\exists x D_1 x), \ldots, T(\exists x D_k x)$.
Let $b_1, \ldots, b_k$ be the first k parameters not in $(\Gamma^{i}_{0})_T \cup \{TB, FC\}$, and put
$\Delta = (\Gamma^{\langle i\rangle}_{\langle 0\rangle})_T \cup \{TB, FC, TD_1(b_1), \ldots, TD_k(b_k)\}$.
Form minimal full Hintikka sets $\Gamma^{\langle i,j\rangle}_{\langle 0,m\rangle}$ from Δ with respect to Par(Δ)$\cup\{b_1\}$ by application of (i)-(viii).

(II) Let the mth split formula in $\Gamma^{\langle i\rangle}_{\langle 0\rangle}$ be of the form $F(\forall x Bx)$, and assume $(\Gamma^{\langle i\rangle}_{\langle 0\rangle})_T \cup \{FB(b)\}$ (b first parameter not in $(\Gamma^{<i>}_{<0>})_T)$ to contain k formulae

$T(\exists xD_1x),\ldots,T(\exists xD_kx)$.

Let $b_1,\ldots,b_k$ be the first k parameters not in $(\Gamma^{<i>}_{<0>})_T \cup\{FB(b)\}$ and let

$\Delta = (\Gamma^{<i>}_{<0>})_T \cup\{FB(b),TD_1(b_1),\ldots,TD_k(b_k)\}$

etc.

Step 1b. $\Gamma^{<i>}_{<0>}$ contains k formulas $T(\exists xD_i(x))$, $1\le i\le k$; let $b_1,\ldots,b_k$ be the first k parameters not occurring in $\Gamma^{<i>}_{<0>}$. Put

$$\Delta = \Gamma^{<i>}_{<0>} \cup\{TD_1(b_1),\ldots,TD_k(b_k)\}.$$

Form all minimal full Hintikka sets $\Gamma^{<i,j>}_{<0,0>}$ by closure of Δ under (i) - (viii) with respect to $\mathrm{Par}(\Delta) \cup \{b_1\}$.

Now step 2 simply repeats this process, giving successors $\Gamma^{<i,j,k>}_{<0,m,n>}$ etc.

Definition. The *search tree* T for the sequent S consists of a tree with the Hintikka sets $\Gamma^{<i0,i1,\ldots,ik>}_{<j0,j1,\ldots,jk>}$ obtained in the search procedure assigned to the nodes; the node with $\Gamma^{\overline{\alpha}k}_{\overline{\varepsilon}k}$ assigned to it has as immediate successors nodes with $\Gamma^{\overline{\alpha}*<u>}_{\overline{\varepsilon}*<v>}$ assigned to it.

The 'Smoothing' of the search tree. To simplify matters, we can show by suitable duplication of nodes that there are i,j (depending on S and x only) such that to each $\Gamma^{\overline{\alpha}x}_{\overline{\varepsilon}x}$ the successors $\Gamma^{\overline{\alpha}x*<u>}_{\overline{\varepsilon}x*<v>}$ are defined for all $u\le i$, $v\le j$. For fixed S, i,j are primitive recursive in x. This can be done e.g. as follows. Assume b,c to have been defined up to x such that for all α,ε with $\forall y<x(\alpha y\le by \;\&\; \varepsilon y\le cy)$ $\Gamma^{\overline{\alpha}x}_{\overline{\varepsilon}x}$ is defined. Let $\mathrm{lth}(n) = \mathrm{lth}(m) = x$, and assume Γ^n_m to contain $l(n,m)$ split formulae. Assume further that the treatment of the pth split formula in Γ^n_m yields $g(p,n,m)$ full Hintikka sets, and that case (b) in the search procedure yields $g(0,n,m)$ full Hintikka sets. We put

$$b(x) = \max\{g(p,n,m)-1 \mid \forall y<x((n)_y\le by \;\&\; (m)_y\le cy \;\&\; p\le l(n,m))\}$$

$$c(x) = \max\{l(n,m) \mid \forall y<x((n)_y\le by \;\&\; (m)_y\le cy)\}$$

and stipulate for $u \le bx$, $v \le cx$

$$(1)\quad \begin{cases} \Gamma^{n*<v>}_{m*<u>} = \Gamma^{n*<g(u,n,m)>}_{m*<u>} & \text{if} \quad u \le l(n,m),\ v \ge g(u,n,m) \\ \Gamma^{n*<v>}_{m*<u>} = \Gamma^{n*<v>}_{m*<l(n,m)>} & \text{if} \quad u > l(n,m) \end{cases}$$

If we wish, we can also make the tree structure of all our search-trees copies of N^N by using (1) for *all* indices u,v, not only for those bounded by $c(x+1)$, $b(x+1)$; at any level only finitely many Γ^n_m are then really relevant.

Note that we may guarantee, if we wish, that $cx > 0$ for all x; the range of the ε in the smoothed search tree then is homeomorphic to a Cantor discontinuum.

7.8. Proof trees and refutation trees.

If we search for a proof of a sequent $S = \Gamma_0$ by means of our search procedure, we must show for *each* $\Gamma^{<i>}_{<0>}$ has a proof; i.e. each $\Gamma^{<i>}_{<0>}$ either contains T (is an axiom) or there is at least one disjunctive possibility (aris ng from treatment of a split formula or taking into account a new parameter in case (b) of the search procedure) such that for some m, all $\Gamma^{<i,j>}_{<0,m>}$ have a proof etc.

In short, a *proof tree* T_α may be coded by a function α such that:

> $\alpha 0 = 0$;
>
> $\alpha 1$ codes the information that for the ith Hintikka set $\Gamma^{<i>}_{<0>}$ we continue with the n_i-th disjunctive possibility; e.g. we may put $\alpha 1 = \langle n_0,\ldots,n_{b0}\rangle$, $n_i \le c1$;
>
> $\alpha 2$ codes the information that for the Hintikka set $\Gamma^{<i,j>}_{<0,k>}$ $(k = (\alpha 1)_i)$ we continue with the mth disjunctive possibility (m depending on i,j), etc.

We note that the proof trees T_α constitute a *fan*, i.e. are represented by a finitely branching tree of functions without finite branches: for $\alpha 1$, the only possibilities are the codings of the mappings (finite in number) from $\{0,\ldots,b0\}$ into $\{0,\ldots,c1\}$; for $\alpha 2$ the only possibilities are codings of mappings from $\{\langle i,j\rangle \mid 0 \le i \le b0 \ \&\ 0 \le j \le b1\}$ into $\{0,\ldots,c2\}$, etc.

Refutation trees may be regarded as 'duals' of proof

trees. If we want to show that the search procedure starting from $S = \Gamma_0$ does *not* lead to a proof, we have to show that for at least one $\Gamma^{<i>}_{<0>}$ *none* of the disjunctive possibilities leads to a proof, i.e. for some i and all k there is a j_k such that $\Gamma^{<i,j_k>}_{<0,k>}$ does not have a proof, etc. The *refutation tree* R_α is coded by α e.g. as follows:

$\alpha 0 = i$ for some $i \le b0$;

$\alpha 1$ codes the information that for the k-th disjunctive possibility associated with $\Gamma^{<i>}_{<0>}$ we continue with the n_k-th set $\Gamma^{<i,n_k>}_{<0,k>}$, $n_k \le b1$, $k \le c1$;

$\alpha 2$ codes the information that for the l-th disjunctive possibility associated with $\Gamma^{<i,n_k>}_{<0,k>}$ we choose the m-th set $\Gamma^{<i,n_k,m>}_{<0,k,l>}$ (m depending on k,l) to continue with, etc.

As before in the case of proof trees, we can show that the collection of refutation trees R_α belonging to a certain $S = \Gamma_0$ constitute a fan. Below we shall tacitly assume α,β,γ to range over finitely branching trees (representing a fan of proof trees or refutation trees) and $\varepsilon,\eta,\varepsilon'$ over (lawless) sequences satisfying $\forall x(\varepsilon x \le cx)$, c as in the preceding section.

A proof tree T_α is said to be *closed* (or to be a *proof* of S) when

$$\forall\beta\exists x(T\in\Gamma^{\overline{\beta}x}_{n} \ \& \ \mathrm{lth}(n)=x \ \& \ n \text{ chosen according to } \alpha).$$

By the fan theorem for β, there is a uniform bound for x; it is easy to see that there is indeed a proof which can be constructed from T_α. Note that we needed the fan theorem with respect to a predicate primitive recursive in α.

A refutation tree R_α is said to be *closed* iff there is a node Γ^n_m (n chosen to correspond to m according to α) which is an axiom. For our completeness theorem the following lemma is crucial:

<u>Lemma.</u> If all refutation trees R_α belonging to $\Gamma_0 = S$ are closed, then S is derivable in the calculus of sequents.

Proof: Let us use $\psi(\alpha,n)$ for the finite sequence m such that Γ^m_n belongs to the refutation tree R_α.
We assume

$$\forall\alpha\exists x\exists n(\mathrm{lth}(n)=x \ \&\ T\bot\in\Gamma_n^{\psi(\alpha,n)}).$$

By the fan theorem applied to a primitive recursive predicate, there is a uniform bound for x and n; weakening the conclusion somewhat, making explicit only the uniform bound for x we obtain

(1) $\exists x\forall\alpha\exists n(\mathrm{lth}(n)=x \ \&\ T\bot\in\Gamma_n^{\psi(\alpha,n)})$

Note that for *given* n, $\psi(\alpha,n)$ ranges over all possible $\overline{\beta}(\mathit{l}\mathrm{th}(n))$. Let us say a proof tree T_α to be a proof of *depth* x if

$$\forall\beta(T\bot\in\Gamma_n^{\overline{\beta}x} \ \&\ \mathrm{lth}(n)=x \ \&\ n \text{ chosen according to } \alpha).$$

Now assume there is no proof of depth x; then for some $\Gamma_{<0>}^{<i>}$ to each k we can find a j such that $\Gamma_{<0,k>}^{<i,j>}$ has no proof of depth x-2, etc. Thus we would obtain that to each n of $\mathrm{lth}(x)$ there was an m of length x such that Γ_n^m was not an axiom; but this contradicts formula (1), and therefore there is a proof of depth x.

7.9. *Refutation trees as generalized Beth models.*

Next we wish to show that any R_α can be interpreted as a generalized Beth model with domain N. Let us write S_n for $\Gamma_n^{\psi(\alpha,n)}$ in a refutation tree R_α. Then

<u>Lemma.</u> Let R_α be a refutation tree for the finite set of signed formulae $S = \Gamma_0$. Let $k,k',k'',\ldots$ range over nodes in the tree; let T be the tree of finite sequences of natural numbers indexing the nodes of R_α, and let $\varepsilon,\varepsilon',\eta$ be used as variables for infinite branches in T, and ξ,ξ' for recursive infinite branches in T. The following properties hold:

(1) $TB,\ FB\in S_k \Rightarrow T\bot\in S_k$

(2) S_k is a full Hintikka set with respect to its free variables;

(3) $TB\in S_k \Rightarrow \underset{\sim}{\forall}k'\geq k(TB\in S_{k'})$ (T-formulae do not get lost in constructing successors in the search tree);

(4) $FP_n(a_{m_1},\ldots,a_{m_{\tau(n)}})\in S_k \Rightarrow$
$\Rightarrow \underset{\sim}{\exists}\xi\in k\underset{\sim}{\forall}k'\in\xi(k'\geq k \Rightarrow FP_n(a_{m_1},\ldots,a_{m_{\tau(n)}})\in S_{k'})$
(this follows from the b-steps in the construction

of the search tree, which do not remove F-formulae);

(5) $F(B \vee C) \in S_k \Rightarrow \underset{\sim}{\exists} \xi \in k \underset{\sim}{\forall} x(\overline{\xi}x \succeq k \Rightarrow FB \in S_{\overline{\xi}x}$ <u>and</u> $FC \in S_{\overline{\xi}x})$

(by the construction of Hintikka sets);

(6) $F(\exists x Bx) \in S_k \Rightarrow \underset{\sim}{\exists} \xi \in k \underset{\sim}{\forall} n \underset{\sim}{\exists} x \underset{\sim}{\forall} y \geq x(FB(a_n) \in S_{\overline{\xi}y})$;

(7) $T(\exists x Bx) \in S_k \Rightarrow \underset{\sim}{\forall} \varepsilon \in k \underset{\sim}{\exists} n \underset{\sim}{\exists} x(TB(a_n) \in S_{\overline{\varepsilon}x})$

(by inspection of the search procedure; in fact, we can assert more: if $T(\exists x Bx) \in S_k$, then $\underset{\sim}{\forall} \varepsilon \in k \underset{\sim}{\exists} n(TB(a_n) \in S_{\overline{\varepsilon}(\mathrm{lth}(k)+1)})$ or equivalently $\underset{\sim}{\forall} y \leq c(\mathrm{lth}(k)) \underset{\sim}{\exists} n(TB(a_n) \in S_{k * \langle y \rangle}))$;

(8) $T(\forall x Bx \in S_k) \Rightarrow \underset{\sim}{\forall} \varepsilon \in k \underset{\sim}{\forall} n \underset{\sim}{\exists} x(TB(a_n) \in S_{\overline{\varepsilon}x})$

(by a succession of b-steps in the search procedure; here also we can in fact assert more, since at level $y+1$ (i.e. for S_k with $\mathrm{lth}(k)=y+1$) at least the first y parameters $a_0,\ldots,a_{y-1}$ have been introduced, so $TB(a_x)$ will certainly appear at $S_{\overline{\varepsilon}(x+2)})$;

(9) $F(B \rightarrow C) \in S_k \Rightarrow \underset{\sim}{\exists} k'(k' \succeq k$ <u>and</u> $TB \in S_{k'}$ <u>and</u> $FC \in S_{k'})$;

(10) $F(\forall x Bx) \in S_k \Rightarrow \underset{\sim}{\exists} k' \underset{\sim}{\exists} x(k' \succeq k$ <u>and</u> $FB(a_x) \in S_{k'})$.

((9) by an a(i)-step. (10) by an a(ii)-step. Here too we can in fact make stronger assertions, e.g. in (9) we could assert

$\ldots \Rightarrow \underset{\sim}{\exists} x(TB \in S_{k * \langle x \rangle}$ <u>and</u> $FC \in S_{k * \langle x \rangle}))$.

Completion of the proof is left to the reader.

<u>Definition.</u> Let R_α be a refutation tree. We define a generalized Beth model $\langle T, \preceq, \phi_\alpha \rangle = M_\alpha$ with $\langle T, \preceq \rangle$ the tree structure underlying R_α, and

$$\phi_\alpha(k) = \{\langle i, n_1, \ldots, n_{\tau(i)} \rangle \mid TP_i(a_{n_1}, \ldots, a_{n_{\tau(i)}}) \in S_k\}$$

Obviously, by (3) of the preceding lemma,

$$k \preceq k' \Rightarrow \phi_\alpha(k) \subset \phi_\alpha(k').$$

<u>Lemma.</u> Let $\langle T, \preceq, \phi_\alpha \rangle$ be the generalized Beth model corresponding to R_α. Then for all $k \in T$, and all $A(a_{n_1},\ldots,a_{n_p})$

(i) $TA(a_{n_1},\ldots,a_{n_p}) \in S_k \Rightarrow k \Vdash A(n_1,\ldots,n_p)$

(ii) $FA(a_{n_1},\ldots,a_{n_p}) \in S_k$ <u>and</u> $k \Vdash A(n_1,\ldots,n_p) \Rightarrow \langle T, \preceq, \phi_\alpha \rangle$ is inconsistent.

Proof: By induction on the logical complexity of $A(a_{n_1},\ldots,a_{n_p})$. We illustrate the basis step.

Assume $TP_i(a_{n_i},\ldots,a_{n_{\tau(i)}})\in S_k$; by the definition of M_α we have $\langle i,n_1,\ldots,n_{\tau(i)}\rangle\in\phi_\alpha(k)$, and therefore $\underset{\sim}{\forall}\varepsilon\in k\underset{\sim}{\exists}x(\bar{\varepsilon}x \Vdash P_i(n_1,\ldots,n_{\tau(i)}))$, i.e. $k \Vdash P_i(n_1,\ldots,n_{\tau(i)})$.

Conversely assume $FP_i(a_{n_1},\ldots,a_{n_{\tau(i)}})\in S_k$ <u>and</u> $\Vdash P_i(n_1,\ldots,n_{\tau(i)})$. Then

(1) $\underset{\sim}{\exists}\xi\in k\underset{\sim}{\forall}k'\in\xi(k'\succeq k \Rightarrow FP_i(a_{n_1},\ldots,a_{n_{\tau(i)}})\in S_{k'})$

and

$$\underset{\sim}{\forall}\varepsilon\in k\underset{\sim}{\exists}x(P_i(n_1,\ldots,n_{\tau(i)})\in\phi_\alpha(\bar{\varepsilon}x) \text{ \underline{or} } M_\alpha \text{ is inconsistent}),$$

i.e.

$$\underset{\sim}{\forall}\varepsilon\in k\underset{\sim}{\exists}x(TP_i(a_{n_1},\ldots,a_{n_{\tau(i)}})\in S_{\bar{\varepsilon}x} \text{ \underline{or} } M_\alpha \text{ is inconsistent}).$$

Using the fan theorem with respect to ε, for a predicate primitive recursive in α, we obtain in the first case a uniform bound x such that

$$\underset{\sim}{\forall}\varepsilon\in n(TP_i(a_{n_1},\ldots,a_{n_{\tau(i)}})\in S_{\bar{\varepsilon}x} \text{ \underline{or} } M_\alpha \text{ is inconsistent}$$

hence also

$$TP_i(a_n,\ldots,a_{n_{\tau(i)}})\in S_{\bar{\xi}x} \text{ \underline{or} } M_\alpha \text{ is inconsistent.}$$

The first possibility contradicts (1), and therefore M_α is inconsistent.

The induction steps, which require an appeal to the lemma in section 7.3, are left to the reader.

<u>Remark.</u> Note that, because of the restriction to generalized Beth models with finitely branching tree structure, we might have avoided an appeal to the fan theorem by simply defining for a model $M = \langle T,\preceq,\phi\rangle$:

$$k \Vdash P_i(n_1,\ldots,n_{\tau(i)}) \equiv_{\text{def}} \underset{\sim}{\exists}x\underset{\sim}{\forall}\varepsilon\in k(\langle i,n_1,\ldots,n_{\tau(i)}\rangle\in\phi(\bar{\varepsilon}x) \text{ \underline{or} } M \text{ is inconsistent})$$

and similarly in the clauses for $\vee$, $\exists$. Of course, the lemma in section 7.3 has to be strengthened correspondingly to: $k \Vdash A \Leftrightarrow \underset{\sim}{\exists}z\underset{\sim}{\forall}\varepsilon\in k(\bar{\varepsilon}z \Vdash k)$.

7.10.

<u>Theorem.</u> (Completeness theorem). Let us assume the Beth models to contain at least the models with $D = N$, and to depend on a choice parameter α ranging over a finitely branching tree $\langle T',\preceq\rangle$.

(i) If the choice sequences satisfy

$$\begin{cases} \forall n\exists\alpha(\alpha\in n) \text{ (n ranging over finite sequences in T')} \\ \forall\alpha\exists xB(\overline{\alpha}x) \to \exists z\forall\alpha\exists x{\le}zB(\overline{\alpha}x) \text{ (B primitive recursive),} \end{cases}$$

then for all sentences A of $\underset{\sim}{\mathrm{IPC}}$

$$\exists x\mathrm{Proof}_{\mathrm{IPC}}(x,\ulcorner A\urcorner) \Leftrightarrow A \text{ valid in all generalized Beth models.}$$

(ii) If the choice sequences satisfy $\forall n\exists\alpha(\alpha\in n)$, then for prenex sentences of $\underset{\sim}{\mathrm{IPC}}$

$$\exists x\ \mathrm{Proof}(x,\ulcorner A\urcorner) \Leftrightarrow A \text{ valid in all Beth models.}$$

As to (ii), note that there is a finite n such that for each refutation tree for $\{FA\}$, A prenex, the tree is either closed at depth n or will remain open if it has not been closed at depth n. This is easily seen by reflection on the structure of the search tree for the case of prenex formulae. Hence, for each generalized Beth model associated with a refutation tree for $\{FA\}$, A prenex, it is decidable whether it is inconsistent or a Beth model.

7.11.

Theorem (Dyson and Kreisel 1961). Under the same assumptions for the range of α as in (i) of the preceding theorem, and in addition

(1) $\forall\alpha\neg\neg\exists xB(\overline{\alpha}x) \to \neg\neg\forall\alpha\exists xB(\overline{\alpha}x)$ (B primitive recursive)

it follows that

$$A \text{ valid in all Beth models } \Rightarrow \neg\neg\exists x\ \mathrm{Proof}_{\mathrm{IPC}}(\ulcorner A\urcorner).$$

Remark. It is not hard to show that (1) is in fact equivalent to the schema (α ranging over N^N, B primitive recursive)

(1') $\forall\alpha\le\lambda x.1\neg\neg\exists xB(\overline{\alpha}x) \to \neg\neg\forall\alpha\le\lambda x.1\exists xB(\overline{\alpha}x)$.

Proof: Let R_A be the following property:

$R_A(\overline{\alpha}x) \equiv_{\mathrm{def}} \overline{\alpha}x$ codes a piece of a refutation tree R_α for $\{FA\}$ which is not closed, i.e. in the nodes Γ^m_n, m chosen according to α, $\mathrm{lth}(m)=x$ there is no occurrence of $T\bot$.

Note that

(2) $\forall xR_A(\overline{\alpha}x) = M_\alpha$ is proper and A is invalid

(since A valid on M_α would mean $\langle\rangle \Vdash A$, conflicting with $FA \in S_{\langle\rangle}$ for proper M_α). Therefore with (2)

$$A \text{ valid on all proper } M_\alpha \Rightarrow \forall\alpha\neg\neg\exists x\neg R_A(\bar{\alpha}x)$$
$$\Rightarrow \neg\neg\forall\alpha\exists x\neg R_A(\bar{\alpha}x)$$

(by (1), α restricted to code functions of refutation trees). $\forall\alpha\exists x\neg R_A(\bar{\alpha}x)$ means that all M_α *are* inconsistent, hence the R_α are closed so $\exists x\ \mathrm{Proof}_{\mathrm{IPC}}(x, \ulcorner A \urcorner)$ and thus

$$A \text{ valid on all Beth-models} \Rightarrow \neg\neg\exists x\ \mathrm{Proof}_{\mathrm{IPC}}(x, \ulcorner A \urcorner).$$

Corollary. Assuming

$$\forall\alpha(\neg\neg\exists x B(\alpha,x) \to \exists x B(\alpha,x))$$

we even obtain

$$A \text{ valid in all Beth-models} \Rightarrow \exists x\ \mathrm{Proof}_{\mathrm{IPC}}(x, \ulcorner A \urcorner).$$

7.12. *Intuitive validity and validity in Beth models.*

As we shall see, there is a direct connection between validity in structures and validity in Beth models, provided we assume the infinite branches (for which we reserve the variables $\varepsilon, \varepsilon', \eta$) of the finitely branching tree underlying any Beth model to be lawless (more precisely, lawless relative to the given finitely branching tree).

Definition. Let us call two Beth models $\langle T, \preceq, \phi\rangle$ and $\langle T, \preceq, \phi'\rangle$ with the same domain D equivalent if ($\Vdash, \Vdash'$ denoting the forcing relations of the respective models)

$$k \Vdash P_i(d_1, \ldots, d_{\tau(i)}) \Leftrightarrow k \Vdash' P_i(d_1, \ldots, d_{\tau(i)})$$

for all $k \in T$, and all i, $d_1, \ldots, d_{\tau(i)} \in D$.

Definition. Let $\langle T, \preceq\rangle$ be a finitely branching tree (of finite sequences of natural numbers), and let $\varepsilon, \varepsilon', \eta$ range over the lawless infinite branches of $\langle T, \preceq\rangle$, k over its nodes.

For any sequence Σ of predicates $P_1^\varepsilon, P_2^\varepsilon, P_3^\varepsilon, \ldots$ (or $\langle P_i^\varepsilon\rangle_i$) we define the Beth model $M(\Sigma) = \langle T, \preceq, \phi\rangle$ by

$$\langle i, d_1, \ldots, d_{\tau(i)}\rangle \in \phi(k) \equiv_{\mathrm{def}} \underset{\sim}{\forall}\varepsilon\in k P_i^\varepsilon(d_1, \ldots, d_{\tau(i)}).$$

Conversely, given a model $M = \langle T, \preceq, \phi\rangle$ we construct a sequence $\Sigma(M) = \langle P_i^\varepsilon\rangle_i$ by

$$P_i^\varepsilon(d_1, \ldots, d_{\tau(i)}) \equiv_{\mathrm{def}} \underset{\sim}{\exists} x(\langle i, d_1, \ldots, d_{\tau(i)}\rangle \in \phi(\bar{\varepsilon}x)).$$

Lemma. Let us assume for the lawless sequences in $\langle T,\leq\rangle$:

$$\begin{cases} \forall k\exists\varepsilon(\varepsilon\in k) \\ A(\varepsilon,\alpha) \to \exists k(\varepsilon\in k \;\&\; \forall\eta\in kA(\eta,\alpha)) \end{cases}$$

where α is a choice parameter for sequences not dependent on lawless sequences in the range of ε, and A not containing non-lawlike parameters besides ε,α. The domain D and its elements may possibly depend on a parameter α, but not on a parameter ε.

(i) The mappings $\Sigma \mapsto M(\Sigma)$, $M \mapsto \Sigma(M)$ are inverse to each other in the following sense: M and $M(\Sigma(M))$ are equivalent models, and Σ, $\Sigma(M(\Sigma))$ are identical (extensionally, as sequences of relations).

(ii) For any $A(P_1,\ldots,P_n)$ of $\underset{\sim}{IPC}$ containing at most the predicate letters $P_1,\ldots,P_n$ we have, for all k

$$k \Vdash A(P_1,\ldots,P_n) \Leftrightarrow \underset{\sim}{\forall}\varepsilon\in k\; A^D(P_1^\varepsilon,\ldots,P_n^\varepsilon)$$

for any Beth model M with forcing relation $\Vdash$ and corresponding sequence $\Sigma(M) = \langle P_i^\varepsilon\rangle_i$.

Proof: (i). Assume $\Sigma = P_1^\varepsilon,P_2^\varepsilon,\ldots$ to be given. Then $\Sigma(M(\Sigma)) = Q_1^\varepsilon,Q_2^\varepsilon,\ldots$ is given by

$$\begin{aligned} Q_i^\varepsilon(d_1,\ldots,d_{\tau(i)}) &\Leftrightarrow \underset{\sim}{\exists}x\underset{\sim}{\forall}\eta\in\bar{\varepsilon}xP_i^\eta(d_1,\ldots,d_{\tau(i)}) \\ &\Leftrightarrow \underset{\sim}{\exists}k[\underset{\sim}{\forall}\eta\in kP_i^\eta(d_1,\ldots,d_{\tau(i)}) \;\&\; \varepsilon\in k] \\ &\Leftrightarrow P_i^\varepsilon(d_1,\ldots,d_{\tau(i)}). \end{aligned}$$

Conversely, assume $\langle T,\leq,\phi\rangle$ to be given and let $M(\Sigma(M)) = \langle T,\leq,\phi'\rangle$. Then

$$\begin{aligned} k \Vdash' P_i(d_1,\ldots,d_{\tau(i)}) &\Leftrightarrow \underset{\sim}{\forall}\varepsilon\in k\underset{\sim}{\exists}x(\langle i,d_1,\ldots,d_{\tau(i)}\rangle\in\phi'(\bar{\varepsilon}x)) \\ &\Leftrightarrow \underset{\sim}{\forall}\varepsilon\in k\underset{\sim}{\exists}x\underset{\sim}{\forall}\eta\in\bar{\varepsilon}xP_i^\eta(d_1,\ldots,d_{\tau(i)}) \text{ (definition of } \phi') \\ &\Leftrightarrow \underset{\sim}{\forall}\varepsilon\in kP_i^\varepsilon(d_1,\ldots,d_{\tau(i)}) \\ &\Leftrightarrow \underset{\sim}{\forall}\varepsilon\in k\underset{\sim}{\exists}x(\langle i,d_1,\ldots,d_{\tau(i)}\rangle\in\phi(\bar{\varepsilon}x)) \\ &\Leftrightarrow k \Vdash P_i(d_1,\ldots,d_{\tau(i)}). \end{aligned}$$

(ii) By induction on the logical complexity of A. For prime formulae, the assertion follows from (i). The induction steps for $A \equiv \forall xBx$, $A \equiv B\&C$ are trivial. Let $A \equiv B \to C$. Then

$$k \Vdash B(P_1,\ldots) \to C(P_1,\ldots) \Leftrightarrow \underset{\sim}{\forall} k' \geq k(k' \Vdash B(P_1,\ldots) \Rightarrow \Rightarrow k' \Vdash C(P_1,\ldots))$$

$$\Leftrightarrow \underset{\sim}{\forall} k' \underset{\sim}{\geq} k(\underset{\sim}{\forall} \varepsilon \in k' B^D(P_1^\varepsilon,\ldots) \Rightarrow \underset{\sim}{\forall} \varepsilon \in k' C^D(P_1^\varepsilon,\ldots))$$

$$\Leftrightarrow \underset{\sim}{\forall} \varepsilon \in k(B^D(P_1^\varepsilon,\ldots) \Rightarrow C^D(P_1^\varepsilon,\ldots)) \Leftrightarrow \underset{\sim}{\forall} \varepsilon \in k A^D(P_1^\varepsilon,\ldots).$$

Let $A \equiv \exists x Bx$; assume $k \Vdash \exists x Bx$. Then for some $d \in D$ $k \Vdash Bd$; hence $\underset{\sim}{\forall} \varepsilon \in k B^D(d,P_1^\varepsilon,\ldots)$, hence $\underset{\sim}{\forall} \varepsilon \in k \underset{\sim}{\exists} d B^D(d,P_1^\varepsilon,\ldots)$ which is equivalent to $\underset{\sim}{\forall} \varepsilon \in k A^D(P_1^\varepsilon,\ldots)$. Conversely, let $\underset{\sim}{\forall} \varepsilon \in k \underset{\sim}{\exists} d B^D(d,P_1^\varepsilon,\ldots), \eta \in k$. Then for some $d \in D$, $B^D(d,P_1^\eta,\ldots)$. For some $k' \geq k$, $\eta \in k'$ and $\underset{\sim}{\forall} \varepsilon \in k' B^D(d,P_1^\varepsilon,\ldots)$, hence $\exists d \in D(k' \Vdash B(d,P_1,\ldots))$; thus we have $\underset{\sim}{\forall} \eta \in k \underset{\sim}{\exists} x \underset{\sim}{\exists} d \in D(\overline{\eta} x \Vdash B(d,P_1,\ldots))$ which implies $k \Vdash \exists x B(x,P_1,\ldots)$.

Remark. On the assumption of the strong definition (as in the remark in section 7.9) for 'k forces A' in the cases for prime formulae, $\exists$ and $\vee$, we need the fan theorem in the proof and the hypothesis has to be correspondingly reinforced assuming $\forall \varepsilon \exists x A(\overline{\varepsilon} x, \alpha) \to \exists z \forall \varepsilon \exists x \leq z A(\overline{\varepsilon} x, \alpha)$.

Theorem. Let us write $\mathrm{Val_B}$ for validity in all Beth models, possibly containing choice parameters α, and based on domains D satisfying the conditions of the lemma; and let Val refer to intuitive validity with respect to domains D, with relations over D depending on choice parameters α, and lawless parameters ε. Let D-Val, D-$\mathrm{Val_B}$ be defined as Val, $\mathrm{Val_B}$ but for a single domain D only. Then

(i) $D\text{-}\mathrm{Val_B}(A) \Leftrightarrow D\text{-}\mathrm{Val}(A)$

(ii) $\mathrm{Val_B}(A) \Leftrightarrow \mathrm{Val}(A)$

(iii) With on the one hand a restriction to lawlike Beth models, and on the other hand a restriction to domains and relations possibly containing a parameter ε, but no α, (i) holds. (Proof immediate by the preceding lemma).

Remarks. (i). It should be noted that the statement of the theorem is not at all plausible for generalized Beth models. Of course, the equivalence $\mathrm{Val_{GB}}(A) \Leftrightarrow \mathrm{Val}(A)$ ($\mathrm{Val_{GB}}$ indicating the validity in all generalized Beth models) holds on th assumption that each generalized Beth model is either a Beth model or inconsistent, but this amounts to acceptance of the excluded third.

(ii). The theorem via the lemma used in an essential way the properties of lawless sequences. Using an appropriate adaptation of the elimination theorem to the case of lawless sequences in a finitely branching tree, we see that validity arguments for Beth models on finitely branching trees can be reduced to arguments not involving infinite lawless branches but only nodes of the tree.

7.13.

<u>Theorem.</u> Let $D = N$, and restrict the relations P_i^* over D to relations recursively enumerable in parameters ε, α. The range of ε consists of lawless sequences in a fan, satisfying (k ranging over finite sequences in the fan)

$$(1) \quad \begin{cases} \forall k \exists \varepsilon (\varepsilon \in k) \\ \forall \varepsilon \exists x A(\overline{\varepsilon} x, \alpha) \rightarrow \exists z \forall \varepsilon \exists x \leq z A(\overline{\varepsilon} x, \alpha) \\ A(\varepsilon, \alpha) \rightarrow \exists k (\varepsilon \in k \ \& \ \forall \eta \in k A(\eta, \alpha)) \end{cases}$$

(cf. our remarks in sections 2.17, 2.18). The range of α consists of choice sequences in a fan; the α's are not dependent on or constructed from the ε's, and are assumed to satisfy (n ranging over finite sequences in the fan)

$$(2) \quad \begin{cases} \forall n \exists \alpha (\alpha \in n) \\ \left.\begin{array}{l} \forall \alpha \exists x A(\overline{\alpha} x) \rightarrow \exists z \forall \alpha \exists x \leq z A(\overline{\alpha} x) \\ \forall \alpha \neg\neg \exists x A(\overline{\alpha} x) \rightarrow \neg\neg \forall \alpha \exists x A(\overline{\alpha} x) \end{array}\right\} \quad (A \text{ primitive recursive}). \end{cases}$$

Then we have weak completeness, i.e.

$$\mathrm{Val}(A) \rightarrow \neg\neg \exists x \ \mathrm{Proof}_{\mathrm{IPC}}(x, \ulcorner A \urcorner).$$

For lawless sequences satisfying (1), and choice sequences satisfying $\forall n \exists \alpha (\alpha \in n)$, we have for prenex A

$$\mathrm{Val}(A) \rightarrow \exists x \ \mathrm{Proof}_{\mathrm{IPC}}(x, \ulcorner A \urcorner).$$

<u>Corollary.</u> (i). Assuming $\forall n \in T \exists \varepsilon (\varepsilon \in n)$, $A\varepsilon \rightarrow \exists n (\varepsilon \in n \ \& \ \forall \eta \in n A \eta)$, and for α as above, using the stronger definition of validity for Beth models with finitely branching trees as underlying partially ordered structures, we have

$$\text{validity in all Beth models} \Rightarrow \neg\neg \exists x \ \mathrm{Proof}_{\mathrm{IPC}}(x, \ulcorner A \urcorner).$$

<u>Corollary.</u> (ii). If we reinforce the version of Markov's schema to

$$\forall \alpha \neg\neg \exists x A(\overline{\alpha} x) \rightarrow \forall \alpha \exists x A(\overline{\alpha} x) \ (A \text{ primitive recursive}),$$

we can replace in the theorem and Corollary (i) '$\neg\neg\exists x\ \mathrm{Proof}_{\mathrm{IPC}}(x, \ulcorner A \urcorner)$' by '$\exists x\ \mathrm{Proof}_{\mathrm{IPC}}(x, \ulcorner A \urcorner)$', i.e. in the first case we obtain completeness instead of weak completeness.

7.14. *Necessary and sufficient condition for weak completeness.*

From Kreisel (1962), van Dalen (1973), we see that conversely weak completeness

$$\mathrm{Valid}(\ulcorner A \urcorner) \rightarrow \neg\neg\exists x\ \mathrm{Proof}_{\mathrm{IPC}}(x, \ulcorner A \urcorner)$$

implies

(1) $\forall\alpha\neg\neg\exists x A(\bar{\alpha}x) \rightarrow \neg\neg\forall\alpha\exists x A(\bar{\alpha}x)$ (A primitive recursive),

which makes the chances of establishing weak completeness minimal. In fact, we can obtain a conflict with Church's thesis CT for *lawlike* functions, provided the range of the α satisfies the fan theorem in the form

(2) $\forall\alpha\exists x A(\bar{\alpha}x) \rightarrow \exists z\forall\alpha\exists x{\le}z A(\bar{\alpha}x)$ (A primitive recursive)

and (a ranging over lawlike sequences)

(3) $\forall\alpha\neg\neg\exists a(\alpha=a)$.

Formula (3) holds for example for choice sequences satisfying $\underset{\sim}{\mathrm{CS}}$ (cf. section 5.6).

Kreisel has shown (see van Dalen (1973) pages 75-86) that on restriction of all domains and relations to lawlike arguments only, and assuming CT, the valid statements of $\underset{\sim}{\mathrm{IPC}}$ are not even recursively enumerable.

7.15.

Remarks. (i). The completeness proof given here (we have chosen a Gödel-type completeness proof for its intuitive appeal, but a Henkin-style proof is possible as well) can easily be extended to theories with a recursively enumerable sets of axioms, if the axioms are enumerated as A_1, A_2, A_3, ..., and we change the search procedure by including at the n-th step TA_n into Δ (in each of the cases aI, aII, b).

(ii) Beth models are directly connected with topological models - see the next section.

(iii) We shall not define Kripke models here (see e.g. Fitting (1969) page 46), but note that a *propositional* Beth model satisfying in addition

(1) $k \Vdash A \vee B \Leftrightarrow k \Vdash A$ <u>or</u> $k \Vdash B$

is a Kripke model. The Beth models in our completeness proof do satisfy (1), and are therefore Kripke models in the propositional case.

(iv) Note that as a byproduct of the completeness proof relative generalized Beth models, we have obtained an intuitionistic model-theoretic proof of closure under cut, where cut is the rule:

$$\text{Cut} \qquad \frac{S_T, FA \qquad S', TA}{S_T \cup S'}$$

To see this, note that the derivability of $S = \{TA_1, \ldots, TA_n, FB_1, \ldots, FB_m\}$ is equivalent to: in each generalized Beth model for which $A_1, \ldots, A_n$ are valid, $B_1 \vee \ldots \vee B_m$ is valid. Assume S_T, FA and S', TA to be derivable. Now consider any generalized Beth model M, and let $S' = S'_T \cup \{FB_1, \ldots, FB_m\}$; if $S'_T \cup S_T$ are valid in M, then A is valid in M, hence $S'_T \cup \{A\}$ is valid in M, and thus $B_1 \vee \ldots \vee B_m$ is valid; it follows that $S_T \cup S'$ is derivable.

7.16. *Beth models as special cases of topological models.*

Given any Beth model $\langle T, \preceq, \phi \rangle$, where now we permit $\langle T, \preceq \rangle$ to be any tree (not necessarily finitely branching) of finite sequences of natural numbers, we may construct a topological model (cf. section 4.17) as follows. Take as topological space Γ the space with the infinite branches of $\langle T, \preceq \rangle$ as points, and the sets $V_k = \{\varepsilon : \varepsilon \in k\}, k \in T$ as a basis for the open sets. Γ is totally disconnected, metrizable; if $\langle T, \preceq \rangle$ is finitely branching, the space is compact, and if $T = N$, i.e. contains *all* finite sequences of natural numbers, Γ is Baire space. Then the assignment

$$(1) \qquad [\![P_i(d_1, \ldots, d_{\tau(i)})]\!] =_{\text{def}} \{\varepsilon : \underset{\sim}{\exists} x (\bar{\varepsilon} x \Vdash P_i(d_1, \ldots, d_{\tau(i)}))\}$$
$$= \{\varepsilon : \underset{\sim}{\exists} x (\langle i, d_1, \ldots, d_{\tau(i)} \rangle \in \phi(\bar{\varepsilon} x))\}$$

yields a topological model with space Γ such that for all sentences A

(2) $\quad [\![A]\!] = \{\varepsilon : \underset{\sim}{\exists} x(\overline{\varepsilon}x \Vdash A)\}$

Conversely, a topological model over Γ gives rise to a Beth model $\langle T, \preceq, \phi\rangle$ if we take

$$\langle i, d, \ldots, d_{\tau(i)}\rangle \in \phi(k) \equiv_{def} V_k \subset [\![P_i(d, \ldots, d_{\tau(i)})]\!],$$

and it is easily seen that this construction is inverse to the construction given by (1).

Note that, in establishing this equivalence, no special assumptions on the range of ε are made; in particular, the correspondence does not depend on ε being lawless.

The correspondence as described above may be extended to topological models for second order theories as described in section 4.17 (cf. van Dalen 1974). The underlying space is Baire space, corresponding to the tree of all finite sequences in the Beth model. The range of the number variables is represented by the natural numbers in the Beth model as well as in the topological model. In the topological model a choice sequence α is represented by a continuous operator (or equivalently, an element of K); in the corresponding Beth model, a choice sequence is represented by a family $\langle\alpha^k\rangle_k$ of partial functions (identified with their graphs) such that

(i) $\quad k \preceq k' \Rightarrow \alpha^k \subset \alpha^{k'}$,

(ii) $\quad$ for any infinite branch ε in $\langle N, \preceq\rangle$, $\{\alpha^{\overline{\varepsilon}x} : x \in N\}$ is total.

We define a 1-1-correspondence which assigns to a family $\langle\alpha^k\rangle_k$ satisfying (i) and (ii) an $e \in K$ such that $\{\alpha^{\overline{\varepsilon}x} : x \in N\} = e|\varepsilon$ and vice versa; this is achieved taking

(3) $\quad e(\hat{x} * n) = y+1 \Leftrightarrow \langle x, y\rangle \in \alpha^n$.

For prime formulae $A \equiv (t[n_1, \ldots, \alpha_1, \ldots] = s[n_1, \ldots, \alpha_1, \ldots])$ we specify the model function ϕ by

$$\langle A, n_1, \ldots, n_p, \langle\beta_1^k\rangle_k, \ldots, \langle\beta_n^k\rangle_k\rangle \in \phi(k') \equiv_{def}$$
$$\exists m \forall i \leq n \forall y < \mathrm{lth}(m)_i \exists z(\langle y, z\rangle \in \beta_i^{k'} \;\&$$
$$\& \; \forall \varepsilon_1 \in (m)_1 \ldots \forall \varepsilon_n \in (m)_n (t[n_1, \ldots, n_p, \varepsilon_1, \ldots, \varepsilon_n] =$$
$$= s[n_1, \ldots, n_p, \varepsilon_1, \ldots, \varepsilon_n])).$$

Now let $e_1, \ldots, e_n$ correspond to $\langle\beta_1^k\rangle_k, \ldots, \langle\beta_n^k\rangle_k$ under (3);

then by induction on the logical complexity of B

$$k' \Vdash B(n_1,\ldots,n_p,\langle\beta_1^k\rangle_k,\ldots,\langle\beta_n^k\rangle_k) \Leftrightarrow$$
$$V_{k'} = \{\varepsilon:\varepsilon\in k'\} \subset [\![B(n_1,\ldots,n_p,e_1,\ldots,e_n)]\!]$$

(verification left to the reader).

7.17. Historical note.

Beth models for intuitionistic predicate logic were developed in Beth (1956), but the (classical and intuitionistic) completeness proofs given there were in need of correction. As a result of criticisms of his proof, Beth gave a revised presentation in section 145 of Beth (1959). Although the details are not very clear, it shows that he realized that for a proper intuitionistic equivalent of the completeness theorem, it was necessary to embed the models in the proper sense in a larger class constituting a fan (called 'semi-models' by Beth); the semi-models may be described suggestively as 'possibly inconsistent' models, or perhaps more accurately, as arising from tentative model constructions which sometimes may turn out to be impossible.

Kripke semantics for intuitionistic predicate logic are developed in Kripke (1965), and a classical Gödel-style completeness proof was given. Henkin-style completeness proofs (classically) were given independently in Aczel (1968), Thomason (1968), Fitting (1969), Luckhardt (1970). W. Veldman rediscovered for Kripke models Beth's idea of the extension to generalized models (containing possibly inconsistent points) and gave an intuitionistic Henkin-style completeness proof based on this idea (Veldman (1974), Lopez-Escobar and Veldman (1975)). H. de Swart then gave a similar Gödel-style completeness proof for Beth models (de Swart 1974). Details as to choice of formal system etc. in this chapter as well as in de Swart's paper are inspired by the treatment in Fitting (1969); the intuitionistic completeness argument here makes use of de Swart's paper. New in this chapter is only a much more detailed and careful analysis of the axioms on choice sequences involved than has been available before, in the light of more recent information on choice sequences.

In Dyson and Kreisel (1961) the original completeness proof of Beth (1956) is analysed and corrected, and it is shown that addition of $\forall\alpha\neg\neg\exists x A(\overline{\alpha}x) \to \neg\neg\forall\alpha\exists x A(\overline{\alpha}x)$ (A primitive recursive) yields weak completeness with respect to Beth-models.

The theorem in section 7.12 on the relationship between Val_B and Val is a generalization of certain results in Kreisel (1958) (dealing with topological models instead of Beth models) and Kreisel (1970) pages 137-139.

APPENDIX A

A SHORT HISTORY OF CHOICE SEQUENCES

A.1.

In this appendix we have attempted to give a brief historical sketch of the development of the theory of choice sequences. The aim is twofold: (i) to provide some materials for the history of intuitionism, (ii) to help the reader to put the (often conflicting) literature on choice sequences into a unified perspective. As to (i), it is well that the reader should realize beforehand the difficulty of presenting a historically correct picture. Where underlying assumptions and concepts were often not explicitly described, one cannot avoid tending to interpret the historical situation in the light of today's insights, and this is bound to create some distortion.

For certain details of a technical nature referred to in this historical account, the reader should consult Appendix B. The notion of a spread law, which plays an important role in traditional intuitionistic literature, and to which we have to refer below, may be defined (at least for our purposes) as follows:

$$\mathrm{Spr}(a) \equiv_{\mathrm{def}} \forall nm(a(n*m) \neq 0 \rightarrow an \neq 0) \\ \& \ \forall n \exists x(an \neq 0 \rightarrow a(n*\langle x\rangle) \neq 0).$$

That is to say, $\{n : an \neq 0\}$ is a (lawlike) countable decidable tree without finite branches (interpreting the variables n,m in the definition as ranging over codes of finite sequences of natural numbers). We define $\zeta \in a$ (the sequence ζ belongs to the spread a) by

$$\zeta \in a \equiv_{\mathrm{def}} \forall x(a(\overline{\zeta}x) \neq 0).$$

A.2. L.E.J. Brouwer.

Choice sequences make their appearance gradually in Brouwer's writings - in Brouwer (1907), which is regarded as the starting point of Brouwerian intuitionism, there is no mention of choice sequences. They first appear, quite unobtrusively, in Brouwer (1912), from which we quote the two relevant passages in the translation by Arnold Dresden (see Brouwer, Collected Works, page 133 (bottom) - p.135 (top)):

'Let us consider the concept: "real number between 0 and 1". For the formalist this concept is equivalent to "elementary series of digits after the decimal point", for the intuitionist it means "law for the construction of an elementary series of digits after the decimal point, built up by means of a finite number of operations". And when the formalist creates the "set of all real numbers between 0 and 1", these words are without meaning for the intuitionist, even whether one thinks of the real numbers of the formalist, determined by elementary series of freely selected digits or of the real numbers of the intuitionist, determined by finite laws of construction.' and two paragraphs later, discussing the continuum hypothesis

'If we restate the question in this form: "Is it impossible to construct infinite sets of real numbers between 0 and 1, whose power is less than that of the continuum, but greater than aleph-null?", then the answer must be in the affirmative; for the intuitionist can only construct denumerable sets of mathematical objects, and if on the basis of the intuition of the linear continuum, he admits elementary series of free selections as elements of construction, then each non-denumerable set constructed by means of it contains a subset of the power of the continuum'.

In these quotations, 'elementary series' is simply a sequence of order type ω. 'Freely selected digits' is the translation of the dutch 'vrije cijferkeuzen' which may also be rendered as 'freely chosen digits' or 'free choices of digits'. In these passages 'formalist' seems to be practically equivalent to 'classical mathematician' (not elsewhere in Brouwer's paper); it is interesting that admitting choice sequences (= free choices of digits) seems to be the formalist's business, not a concern of the intuitionist, who deals only with lawlike sequences; the second quotation however shows that the intuitionist at least conceivably might use choice sequences.

Between 1912 and 1914 Brouwer seems to have changed his mind[†] as to the admissibility of choice sequences in intuitionistic mathematics - at least in his review of the book by Schoenflies and Hahn, 'Die Entwickelung der Mengenlehre und ihrer Anwendungen' (Brouwer 1914) he expresses himself as follows (second paragraph, translated from the German):

'To illustrate this point more clearly, I recall that for the intuitionist only such infinite sets exist which are composed from ..... and a second part, which is based on ..... while each of its elements is determined from a sequence of choices of elements from a finite or countably infinite set in such a way that ..... '.

Still later, Brouwer (in Brouwer 1918) expresses himself quite clearly: on page 3, second paragraph, a spread is introduced as follows (in free translation):

'A spread (German:Menge) is a law (German:Gesetz) such that to each choice in a sequence of successive choices of natural numbers the law assigns either a definite sign (German:Zeichen) or nothing, or ..... . Each sequence which is produced in this manner (hence in general not representable as a completed object) is said to be an *element of the spread.*

When distinct choice sequences (German:Wahlfolgen) always produce distinct sequences of signs, the spread is said to be *individualized*'. Concepts which are being defined are in italics in Brouwer (1918); but this appearance of 'choice sequence' is not in italics.

It is worth noting that in Brouwer (1919), the sequel

[†] I owe the references to Brouwer (1912) and (1914) to Prof. Dr. A. Heyting, who also stressed their relevance for showing the development of Brouwer's ideas.

to Brouwer (1918), there is an application of 'König's lemma' (or rather the fan theorem without continuity assumptions: a finitely branching tree all of whose branches are finite is itself finite) in the proof of the theorem on page 7, lines 12-16, which there is obviously regarded as evident.

A further development is found in Brouwer (1923), where for the first time the well-known theorem of intuitionistic analysis is stated: each real-valued function defined on [0,1] is uniformly continuous.

For an operator Φ defined on all choice sequences of a spread, it is assumed (as being evident) that such an operator is always defined by a sequence of operators, $\Phi_1, \Phi_2, \ldots, \Phi_n$ such that Φ_n as a constant determines the modulus of continuity needed for computing Φ_{n-1}, and $\Phi_{n-1}\alpha$ determines the modulus needed for computing $\Phi_{n-2}\alpha$, etc. An assumption amounting to 'König's lemma' is again tacitly used. Obviously, Brouwer must have realized that this treatment was unsatisfactory, and thus in Brouwer (1924, 1924A) we find the first version of the much discussed proof of the 'bar theorem'. The continuity assumptions implicit in the discussion seem to point to $\forall\alpha\exists!x$-continuity only, not to $\forall\alpha\exists x$-continuity: there is mention only of a *law* which assigns a natural number to each element of a spread. If this (as would be in keeping with the *usual* interpretation of 'one-valued function') is interpreted extensionally, the assumptions made about the existence of the species μ_1 in Brouwer (1924) amount to $\forall\alpha\exists!x$-continuity. With little change, the argument is again produced in Brouwer (1927).

In Brouwer (1925), for the first time Brouwer felt the need to be more explicit about the concept of a choice sequence; he repeats (on page 245) the definition of 'element of a spread': 'Each sequence of signs[3], generated in this manner by an unlimited choice sequence ..... is said to be an element of the spread'. The footnote 3 elaborates this by the remark:

'Including the feature of their freedom of continuation, which after each choice can be limited arbitrarily (possibly to being fully determined, but anyway according to a spread law)'

This seems to point to a concept as described in Example IV of Appendix B. As shown in section B4, it is indeed possible to justify $\forall\alpha\exists!x$-continuity for this notion, but at the same time it does not have all the properties Brouwer must have tacitly assumed for it: for example, the sequences of Example IV cannot be proved to be closed under very simple operations which occur in analysis, such as $\lambda\alpha\lambda x.2\alpha x$. To see this intuitively, one observes that a relation such as $\beta = \lambda x.2\alpha x$ not only imposes a relation between the lawlike spreads to which α and β can belong, but also it provides information about β going beyond 'β is lawless relative to a given spread', namely the fact that β is completely determined relative to another sequence α belonging to a given spread. For a detailed discussion see Troelstra (1969B).

Brouwer's elaborations apparently rather stem from a desire to present a clearer picture of how one should think of a choice sequence, than from a wish to describe a notion for which certain principles could then be *shown* (informally, but rigorously) to be valid.

Brouwer returns to the concept of choice sequence on two later occasions: Brouwer (1942), under section 2, first liberalizes[†] the interpretation of the footnote quoted above,

---

[†]This more liberal interpretation is already indicated in a handwritten correction by Brouwer in his copies of Brouwer (1925) (see editors note ⟦3⟧ on page 590 of Brouwer, Collected Works), which adds to the footnote 3:

'The arbitrariness of this possible restriction, which does not violate the possibility of continuation, adds a new element of arbitrariness to this choice sequence and its continuations. It is also possible to add a well-ordered species of restrictions in the spread (e.g. a restriction of the existing freedom of adding restrictions on future choices)'.

by adding:

'Moreover, the freedom of continuation of a sequence of signs generated by an unlimited sequence of choices, which represents a spread element, may be restricted arbitrarily after each choice (e.g. up to complete determinateness, or corresponding to a spread law); in fact, the arbitrariness of these restrictions to be assigned to the choices (which must always leave open the possibility for continuation) represents an essential characteristic of the free generation of the spread element. To each individual restriction we can further assign a second order restriction, which restricts the arbitrariness of further restrictions, etc.'.

This statement is then again partially retracted in Brouwer (1952) page 142, in a footnote:

'In former publications I have sometimes admitted restrictions of freedom with regard also to future restrictions of freedom. However, this admission is not justified by close introspection, and moreover would endanger the simplicity and rigour of further developments'.
(This statement has sometimes been interpreted as excluding lawless sequences, since imposing no restrictions (of first order) is itself a second-order restriction - but when all statements of Brouwer on the subject are compared, this seems to be a quite unnatural interpretation, especially in view of the phrase 'endanger the simplicity...' - although Brouwer may have failed to recognize the simplicity of the notion of lawless sequence.)

A.3. The first axiomatizations: Heyting, Kleene.

A hesitant beginning with the formalization of the theory of choice sequences is made in Heyting (1930); a form of the continuity theorem is stated, but not of the bar theorem. The subject was taken up again in Kleene (1957), researches culminating in the axiomatization as presented in Kleene and Vesley (1965). In this monograph, little or nothing in the way of conceptual analysis is attempted;

describing intuitionistic practice was a principal aim, and the axiom schemas are presented as a codification of the principles implicitly adopted by Brouwer. $\forall\alpha\exists\beta$-continuity is adopted with the remark (Kleene and Vesley 1965), page 73):

'Still, it seems to us that the intuitionistic reasons for accepting the principle for numbers apply equally to the principle for functions (even though we do not know of an explicit affirmation of it in Brouwer's writings).

In fact, as we have remarked above, Brouwer's papers seem to imply an assumption of $\forall\alpha\exists!x$-continuity only, not $\forall\alpha\exists x$-continuity as assumed by Kleene. Our contention that Brouwer postulated $\forall\alpha\exists!x$-continuity depends on the assumption that an operator assigning a natural number to each α would automatically be interpreted as an extensional operator by the mathematicians of that period.

A.4. Lawless sequences.

The concept of lawless sequence ('absolutely free choice sequence') first appears in Kreisel (1958), for sequences belonging to a finitary spread; its section 9 states the elimination theorems for formulae containing at most one lawless parameter. In Kreisel (1965), the lawless sequences are briefly discussed, and their connection with forcing is pointed out (pages 110, 142). The full theory $\underset{\sim}{\mathrm{LS}}$ is first described in Kreisel (1968), except that axiom 2.4 as stated there, is incorrect and should be replaced by LS4 as in our Chapter 2. The elimination theorems are stated and their proofs sketched. As it stands, Kreisel's description of the elimination mapping for the theory *including* species variables and comprehension for species is adequate for the first half of the elimination theorem (our section 3.14), but not for the second half (our section 3.15; for an elimination mapping which is also adequate for the second half of the elimination theorem in the presence of species variables, see sections 3.16 - 3.18).

Lawless sequences are also briefly discussed in Myhill

(1967), Troelstra (1968); fairly detailed expositions are in Troelstra (1969) and in a more general context in Troelstra (1969B). Projections of lawless sequences are first used in Kreisel (1967) (see e.g. pages 180-182), and again in Kreisel (1968). The interest in projections constructed from lawless sequences as a 'universe of sequences for analysis' derives from the fact that for lawless sequences we *do* have an (essentially) simple and convincing conceptual analysis (which, however, was discovered by degrees).

In section 2.1 we have already listed the main reasons that justify our interest in lawless sequences. It should be added here, that they have been extremely useful also in connection with the 'machinery' of $\mathbf{IDB}_1$ (described in section 3.1) which was originally developed for the purpose of eliminating lawless sequences, and could be transferred without essential changes to the case of **CS**.

A.5. **FIM**, *the missing link with respect to elimination of choice sequences, and* **CS**.

The system **FIM** of Kleene and Vesley (1965) as such is not suited to an elimination ('contextual definition') of choice quantifiers since no other quantifiers are present (as to the reasons for not including quantifiers over lawlike sequences in **FIM**, see Kleene (1968)). But if we extend **FIM** in the more or less obvious way by addition of variables and quantifiers for lawlike sequences, we see that the resulting system (say **FIM***) provides an important step forward (when compared with Brouwer) towards elimination of choice sequences since it contains, in addition to principles implicit in Brouwer's writings, $\forall\alpha\exists\beta$-continuity.

But one important link was still missing: a schema permitting the treatment of implication in the definition of the elimination mapping. The type of schema needed was well known from the theory of lawless sequences, where the axiom schema of open data (LS3) took care of implication. LS3 in its simplest form can be stated (cf. section 2.6) as (α the only lawless parameter in A)

(1) $A\alpha \to \exists n(\alpha \in n \ \& \ \forall \beta \in n A\beta)$.

This suggested extending (something like) FIM* by a schema of the form

(2) $A\alpha \to \exists R \in \mathcal{R}(R\alpha \ \& \ \forall \beta(R\beta \to A\beta))$

where $\mathcal{R}$ is a class of lawlike predicates of sequences. Formula (2), which is a generalization of (1), permits us to prove

$$\forall \alpha(A\alpha \to B\alpha) \leftrightarrow \forall R \in \mathcal{R}(\forall \alpha \in R(A\alpha) \to \forall \alpha \in R(B\alpha)),$$

which enables us to deal with implication.

Since choice sequences in the intuitionistic literature always appeared in the context of a spread, Kreisel proposed[†] to express 'taken from a spread' by the following special form of (2)

(3) $A\alpha \to \exists a[\mathrm{Spr}(a) \ \& \ \alpha \in a \ \& \ \forall \beta \in a(A\beta)]$.

This schema was regarded as justified by the fact that a choice sequence was a priori given as belonging to a certain spread.

Neither Kleene and Vesley (1965), nor Kreisel (1963) paid special attention to closure of choice sequences under continuous operations (which was implied by the postulates). However, closure under continuous operations is not intuitively valid for a concept of sequence 'lawless relative to some lawlike spread', which had been proposed by Kreisel as a model of (3). In fact (3) together with closure under continuous operations permitted an easy refutation of Church's thesis as shown in Troelstra (1968), whereas (assuming the elimination theorems to hold) Kreisel's system should be consistent with Church's thesis.

The oversight was repaired by substitution of the principle of analytic data for (3); the resulting system was first described in Troelstra (1968). Detailed proofs of the elimination theorem were presented in Kreisel and

---

[†]According to Kreisel (in correspondence), when writing Kreisel (1963) he was not aware of the relevant footnote in Brouwer (1925).

Troelstra (1970); there the original aim of Kreisel (1963), reduction of bar-induction to Π^1_1-comprehension was achieved. It should be added that the discovery of the oversight certainly stimulated a more careful and 'informally rigorous' description of concepts. Kreisel (1963) has not been officially published, but an account of the system may be found, in Kreisel (1965), 2.5, 2.6 (pages 133 - 143).

A.6. Myhill: the rôle of intensional aspects of choice sequences.

J. Myhill, in Myhill (1967), took up the original idea of Brouwer as described in Brouwer (1925) (see example IV). It seems plausible that Kreisel (1967) was another source of inspiration.

Myhill presents reasons for accepting $\forall\alpha\exists x$-continuity, but not $\forall\alpha\exists\beta$-continuity. The paper shows a clear awareness of the intensional aspects of choice sequences and the possibility that the value of an operator applied to a choice sequence may depend on the intensional aspects too.

Myhill also notes that (page 292) we are not permitted to make use† of (Δ) : $\Pi_1(\mathrm{Abstr}(\chi))x \equiv U$ for all x (notation of B3), since (Δ) would imply that a type of information about Abstr(χ) would be available, which is incompatible with our intention to regard Abstr(χ) as the same sort of sequence as χ (for χ of the sort described in Example IV).

Myhill's derivation of $\forall\alpha\exists x$-continuity is still defective: a more careful discussion yields at most $\forall\alpha\exists !x$-continuity (see section B4) whereas, in general, $\forall\alpha\exists x$-continuity can only be expected to hold under suitable restrictions on the predicates to which the schema is applic-

---

†Myhill does not use 'Abstr'; this was introduced in Troelstra (1968), and discussed at greater length in Troelstra (1969, 1969B) and the present book, in order to make Myhill's discussion more perspicuous and to demonstrate the essential point better.

able (cf. the results for projections of lawless sequences, discussed in section B7). Myhill's counterexample to $\forall\alpha\exists\beta$-continuity is also not conclusive. The counterexample is based on acceptance of a schema (usually named 'Kripke's schema' in the literature) which (in a somewhat strengthened form) may be rendered as

KS $\quad \exists\xi[\exists x(\xi x=0) \leftrightarrow A]$

for any predicate A, ξ a variable over number-theoretic sequences. The schema is really inspired by the theory of the 'creative subject', not to be discussed in these notes (for an exposition see Troelstra (1969), section 16).

Myhill's counterexample is essentially as follows: from KS, we obtain, taking for A a property with choice parameter α (namely $\forall x(\alpha x = 0)$):

(3) $\quad \forall\alpha\exists\xi[\exists x(\xi x=0) \leftrightarrow \forall x(\alpha x=0)]$

which is read by Myhill as

(4) $\quad \forall\alpha\exists\beta[\exists x(\beta x=0) \leftrightarrow \forall x(\alpha x=0)]$

(i.e. the ξ is supposed to range over choice sequences of the kind considered by Myhill). Now $\forall\alpha\exists\beta$-continuity would require the β to be determined continuously from α, but this is obviously impossible. (To see this, consider the β belonging to $\alpha=\lambda x.0$; then obviously for some x_0, βx_0 is required; but βx_0 would depend on an initial segment of α only, which is contradictory). However, the crux of the matter is indeed in the assumption that the ξ may be interpreted as a β, i.e. as a choice sequence of the universe considered; but this would only be obvious if we could be assured that this universe would also be closed under non-continuous operations such as are implicit in (3); in short, we have only proved (3), not (4); and no obvious way of containing a counterexample suggests itself which is not open to this kind of criticism. We shall tacitly bypass Myhill (1968, 1970) in this review, since they offer no points of interest for the present survey, and moreover do not seem to be based on any concept of choice sequence in particular.

A.7. Choice sequences satisfying FIM *and* CS.

In Troelstra (1968), there is a not very successful attempt to justify $\forall\alpha\exists\beta$-continuity for a certain notion of sequence (invented with the express purpose of validating CS) by introduction of a theory of many creative subjects, an idea which was again abandoned in Troelstra (1969). The notion of choice sequence proposed for CS (discussed in Appendix C), is not fully convincing as an example of conceptual analysis - it is too complicated. As will be clear from the preceding historical discussion, this is not astonishing - formal criteria led to a change of the system proposed in Kreisel (1963), and then afterwards the notion was invented in an attempt to validate CS. Cf. also section 5.1. For these reasons we relegated the discussion of this notion to Appendix C.

APPENDIX B

INTENSIONAL ASPECTS OF CHOICE SEQUENCES IN CONNECTION WITH CONTINUITY AXIOMS

B.1.

In this section we discuss by means of some examples how 'intensional aspects' (in the sense of section 1.4) influence the validity and non-validity of continuity postulates. We have three aims in doing so: (i) to provide some illustrations to claims made in the preceding Appendix; (ii) to serve as an introduction to Appendix C; (iii) to provide some further examples of how new universes of sequences may be constructed from lawless sequences (see especially section B7).

B.2.

Example I. A first crude attempt to combine the advantages of lawless sequences with the desirability of having all lawlike sequences in the universe would be to take simply the union of lawlike sequences and lawless sequences as our universe. But then continuity does not hold, nor even for the quantifier combination $\forall\alpha\exists!x$, as is seen by the following counterexample. Take for $A(\alpha,x)$

$$A(\alpha,x)\equiv[(x=0\ \&\ \exists a(\alpha=a))\vee(x=1\ \&\ \neg\exists a(\alpha=a))].$$

Obviously, since every element of our universe is unambiguously either lawless or lawlike, $\forall\alpha\exists!xA(\alpha,x)$; but just as clearly x cannot be found from an initial segment of α. Now, noting that in defining A we had to make use of variables for lawless sequences, it seems natural to ask whether perhaps $\forall\alpha\exists!x$-continuity would hold when restricting the language to $\underset{\sim}{\mathrm{EL}}$. But the answer is no: take for $A^*(\alpha,x)$

$$A^*(\alpha,x)\equiv[(x=0\ \&\ \exists\beta\forall x(2\alpha x=\beta x))\vee(x=1\ \&\ \neg\exists\beta\forall x(2\alpha x=\beta x))]$$

which is $A(\alpha,x)$ in disguise, since the lawlike sequences within the universe may be characterized by the fact that 'doubling their values' is a continuous operation not leading outside the universe, whereas for lawless sequences this operation yields a sequence which is neither lawlike nor lawless.

Bar induction on the other hand *does* hold, since we can prove the (equivalent) principle SBC

SBC $\quad \forall\alpha\exists x A(\overline{\alpha}x) \rightarrow \exists e \forall\alpha A(\overline{\alpha}(e(\alpha)))$

(cf. Kreisel and Troelstra (1970), 5.6.1, 5.6.3). To see this, assume $\forall\alpha\exists x A(\overline{\alpha}x)$; then it follows that (using ε,η as variables for lawless sequences) $\forall\varepsilon\exists x A(\overline{\varepsilon}x)$, hence $\forall\varepsilon A(\overline{\varepsilon}(e(\varepsilon)))$ for some e. Without restriction we may assume $en=y+1 \rightarrow \mathrm{lth}(n)\geq y$ for e; now take any α of our universe, let $\alpha\in n, en=y+1$; then there is an $\varepsilon\in n$, and $A(\overline{\varepsilon}(e(\varepsilon)))$ follows. But since $\overline{\varepsilon}(e(\varepsilon))\preceq n$, it follows with $e(\varepsilon)=e(\alpha)$ that $\overline{\varepsilon}(e(\varepsilon))=\overline{\alpha}(e(\alpha))$, and therefore $A(\overline{\alpha}(e(\alpha)))$.

B.3. Description of a general framework for the other examples.

For the description of some further examples, let us first indicate a certain generalization of the concept of a lawless sequence. To set the stage, think of sequences of the following type: at each stage we select a value x_n and a restriction R_n on future choices, i.e. we choose pairs

$$\langle x_0,R_0\rangle,\langle x_1,R_1\rangle,\langle x_2,R_2\rangle,\ldots$$

of numerical values and restrictions on future choices, the restrictions becoming narrower (although perhaps not properly so) as we proceed. In principle, we might also consider the possibility of second and higher order restrictions, i.e. restrictions on restrictions etc. We shall not do so here apart from assuming certain *general* and *uniform* second order restrictions for each notion of sequence considered (i.e. the restrictions specifying which restrictions of first order R_i are permitted, and which R_{i+1} can be chosen after the choices already made).

For a convenient description, let us introduce some notation. Let us use ξ,χ,η for sequences of pairs of numbers and restrictions, i.e. ξ is of the form

$\langle x_0,R_0\rangle, \langle x_1,R_1\rangle, \langle x_2,R_2\rangle,\ldots,$ and $\xi n = \langle x_n,R_n\rangle$. We define

$$\Pi_0(\xi n) \equiv x_n,\ \Pi_0\xi \equiv \langle x_n\rangle_n,$$
$$\Pi_1(\xi n) \equiv R_n,\ \Pi_1\xi \equiv \langle R_n\rangle_n.$$

The purpose of introducing this notation is to make it possible to discuss in a more *explicit* way intensional aspects of the sequences we are dealing with; so far we did not show intensional aspects in our notation, they were implicit. However, the mathematically interesting properties of such sequences may be supposed to depend on the numerical values only, i.e.

$$A(\Pi_0\xi,x)\ \&\ \Pi_0\xi=\Pi_0\eta \rightarrow A(\Pi_0\eta,x).$$

It is intended of course that $\Pi_0\xi$ should satisfy all R_n, which can be expressed by $\forall n(\Pi_1\xi n(\Pi_0\xi))$. Now to make the description more definite, we further stipulate:

(i) The R_n are assumed to belong to a decidable class $\mathcal{R}_n$; $\cup\mathcal{R}_n=\mathcal{R}$. 'Decidable' means that an element of $\mathcal{R}_n$ can be seen to belong to $\mathcal{R}_n$ from the way it is given to us - no separate proof is needed.

(ii) A decidable condition $\sqsubset$ is given, and $\Pi_1\xi$ should satisfy $\forall x(\Pi_1\xi(x+1)\sqsubset\Pi_1(\xi x))$, and $R_{n+1}\sqsubset R_n' \rightarrow R_{n+1}\cap W\subset R_n'\cap W$ (W being a set of sequences having a common initial segment of length $n+2$ consistent with R_{n+1}).

(iii) The elements of $\mathcal{R}$ are assumed to be lawlike (completely determined). Moreover, we shall assume it to be decidable whether $\chi(n+1)$ is consistent with R_n, i.e. can be extended to χ such that $R_n(\Pi_0\chi)$; and of course future choices must be consistent with R_n.

(iv) To satisfy the demands under (iii) in a perspicuous way, we assume $R_n\chi$ to be of the form $\forall x(R_n^*(\overline{\chi}(x+1)))$, with R_n^* decidable, and such that $R_{n+1}\sqsubset R_n \rightarrow (R_{n+1}^*(\overline{\chi}x) \rightarrow R_n^*(\overline{\chi}x))$ for $x > n+1$. Any admissible initial segment should be extendible (for a detailed description of the necessary stipulations see either Troelstra (1969B) or van Dalen and Troelstra (1970)).

Now relative to the specification of $\mathcal{R}$ and $\sqsubset$ we may think of our sequences as analogues of lawless sequences: at any stage only a finite initial segment $\langle x_0,R_1\rangle,\ldots,\langle x_n,R_n\rangle$

is known; we make no restrictions on future choices except as specified by the given $\mathcal{R}$ and $\sqsubset$. The reason for insisting on the decidability of $\mathcal{R}_i$ and $\sqsubset$ is in the fact that if we do not require decidability of $\mathcal{R}_i$, $\sqsubset$, then operations acting on initial segments $\langle x_0, R_0\rangle, \ldots, \langle x_n, R_n\rangle$ might conceivably also use information not made explicit[†]: i.e. a *proof* of the admissibility of this initial segment (including proofs of $R_i \in \mathcal{R}_i$, $R_{j+1} \sqsubset R_j$ for $0 \leq i \leq n$, $0 \leq j \leq n$). Assuming decidability however permits us to restrict attention to the explicitly given 'intensional' information, since decidability entails the existence of *standard* proofs of admissibility of an initial segment, determined by the initial segment itself only.

B.4.

<u>Example II.</u> Using ζ as a variable for arbitrary sequences of type $N \to N$, we define the conditions

$$U \equiv \lambda\zeta.[0=0],$$
$$N_x \equiv \lambda\zeta.[\forall z \geq x(\zeta z = 0)],$$
$$B_x \equiv \lambda\zeta.[\forall z \geq x(\zeta z = 1)].$$

So U is the 'universal' condition (i.e. no restriction). For this example we put

$$\mathcal{R}_x = \{U, N_x, B_x\},$$

and holds in the following cases:

$$U \sqsubset U, N_x \sqsubset U, B_x \sqsubset U;$$
$$B_{x+1} \sqsubset B_x, N_{x+1} \sqsubset B_x;$$
$$N_{x+1} \sqsubset N_x.$$

We show informally that for this example we cannot expect $\forall\alpha\exists x$-continuity to hold. Take for $A(\alpha, x)$

[†]The point is not very substantial however; although it is necessary to make it as a matter of hygiene, we do not know of any examples where these possibilities play a rôle. On the other hand, the condition is fulfilled in our examples.

$$A(\alpha,x) \equiv \exists\chi[\alpha=\Pi_0\chi \ \& \ \{(\Pi_1\chi 0\equiv U \ \& \ x=0) \vee (\Pi_1\chi 0\equiv B_0 \ \& \ x=1) \vee (\Pi_1\chi 0\equiv N_0 \ \& \ x=2)\}];$$

α may be taken to range over sequences $\Pi_0\chi$, χ lawless relative $\mathcal{R}_x$, $\sqsubset$ as stipulated.

Obviously

(1) $\forall\alpha\exists x A(\alpha,x)$

and $A(\alpha,x) \ \& \ \alpha=\beta \rightarrow A(\beta,x)$ (i.e. A is extensional).

We wish to show that for no K-function e

(2) $\forall\alpha A(e,e(\alpha))$

holds. Assume (2) to hold for some e, and determine $n=(\overline{\lambda x.0})y$ such that $en\neq 0$; we may assume $y>1$.

Case a. $en=3$. Then also for χ starting with $\langle 0,U\rangle,\ldots,\langle 0,U\rangle$ of length y $e(\Pi_0\chi)=3$, so there must be a ξ such that $\Pi_0\chi=\Pi_0\xi$, $\xi 0\equiv\langle 0,N_0\rangle$; so $\Pi_0\chi=\Pi_0\xi=\lambda z.0$. This holds for any such χ starting with $\langle 0,U\rangle,\ldots,\langle 0,U\rangle$ of length y, conflicting with the fact that we may choose χy to be $\langle 1,U\rangle$. Therefore $en\neq 3$.

Case b. $en=2$. In the same manner, for any χ starting with $\langle 0,U\rangle,\ldots,\langle 0,U\rangle$ of length y there is a ξ with $\Pi_0\xi=\Pi_0\chi$, $\xi 0\equiv\langle 0,B_0\rangle$; hence $\forall x(\Pi_0\chi x=\Pi_0\xi x\leq 1)$, conflicting with the possible continuation $\langle 0,U\rangle,\ldots,\langle 0,U\rangle,\langle 2,U\rangle$ for χ; hence $en\neq 2$.

Case c. $en=1$. Then to any χ starting with $\langle 0,R_0\rangle,\ldots,\langle 0,R_{y-1}\rangle$ we should be able to find a ξ with $\Pi_0\chi=\Pi_0\xi$, such that $\xi 0=\langle 0,U\rangle$.

$\Pi_0\xi=\Pi_0\chi$ must be known from certain initial segments σ,τ of χ,ξ respectively, say of length $x\geq y$. But if $R_0\equiv U$, then ξ,χ must be thought of as independent processes, so that $\Pi_0\xi=\Pi_0\chi$ can only be known if σ,τ both end with restrictions N_{x-1}; i.e. χ,ξ are both lawlike. But it is intuitively absurd to assume that we would know a priori that all sequences starting with $\langle 0,B_0\rangle,\ldots,\langle 0,B_{y-1}\rangle$ are lawlike (as to their numerical components). Hence also $en\neq 1$, and thus a contradiction has been obtained from (2). An indication of how to put this argument on a more rigorous basis by assuming suitable axioms is given in B.8.

On the other hand, for this example (and more generally for similar concepts of sequence satisfying certain conditions) we are able to show the validity of the $\forall\alpha\exists!x$-continuity schema (WC-N) i.e.

$$\forall\alpha\exists!xA(\alpha,x) \to \forall\alpha\exists x\exists y\forall\beta(\overline{\alpha}x=\overline{\beta}y\to A(\beta,y)).$$

To justify this schema, we shall assume about our concept of sequence

(a) $U\in\mathcal{R}_x$ for all x, $U\sqsubset U$, and for any $R_x\in\mathcal{R}_x, R_x\sqsubset U$.

(b) Let $\langle x_0,R_0\rangle,\ldots,\langle x_n,R_n\rangle$ be an admissible initial segment, then there is an a with $a(n+1)=\langle x_0,\ldots,x_n\rangle$ such that $\langle x_0,R_0\rangle,\ldots,\langle x_n,R_n\rangle$, $\langle a(n+1),R^{(a)}\rangle$ is admissible, where $R^{(a)}\equiv\lambda\zeta.[\zeta=a]$.

(c) Condition (a) permits the introduction of a process 'Abstr' similar to the one used in the discussion of continuity postulates for lawless sequences; if ϕ is an operation mapping the sequences considered into the natural numbers, then $\phi(\text{Abstr}(\chi))$ is determined by the action of a neighbourhood function e on $\Pi_0\chi$ which we assume to be total, i.e.

$$\forall nm(en\neq 0 \to en=e(n*m))$$

$$\forall\zeta\exists x(e(\overline{\zeta}x)=0)$$

where as before ζ ranges over *all* sequences of natural numbers.[†]

(a) and (b) are satisfied in our example; (c) amounts to an intuitively justifiable postulate.

Assume $\forall\alpha\exists!xA(\alpha,x)$ (α ranging over sequences $\Pi_0\chi$); then by a selection principle there is an operator Φ such that $\forall\chi A(\Pi_0\chi,\Phi\chi)$ with $\Pi_0\chi=\Pi_0\xi \to \Phi\chi=\Phi\xi$. Hence also

$$\forall\chi A(\Pi_0\text{Abstr}(\chi),e(\Pi_0\chi)).$$

Now assume for some χ, $\Phi\chi$ to be determined from $\langle x_0,R_0\rangle,\ldots,\langle x_n,R_n\rangle$, and $\Phi(\text{Abstr}(\chi))$ to be determined from $\langle x_0,U\rangle,\ldots,\langle x_n,U\rangle$; without restriction we may assume $n\geq m$.

[†]This does not commit us to regarding the quantification over all sequences of natural numbers as meaningful for statements of arbitrary complexity!

Let $\chi(n+1)=\langle x_{n+1},R_{n+1}\rangle$, and consider a ξ starting with

$$\langle x_0,R_0\rangle,\ldots,\langle x_n,R_n\rangle,\langle x_{n+1},R^{(a)}\rangle \quad (x_{n+1}=a(n+1))$$

and an η starting with

$$\langle x_0,U\rangle,\ldots,\langle x_n,U\rangle,\langle x_{n+1},R^{(a)}\rangle.$$

Now

$$A(\Pi_0\mathrm{Abstr}(\chi),e(\Pi_0\chi)),\ \Phi(\mathrm{Abstr}(\chi)) = e(\Pi_0\chi),$$

hence

$$A(\Pi_0\eta,e(\Pi_0\chi)), \qquad \Phi(\eta) = e(\Pi_0\chi),$$

also

$$A(\Pi_0\chi,\Phi\chi)$$

and

$$A(\Pi_0\xi,\Phi\chi).$$

Since $\Pi_0\xi=\Pi_0\eta=a$, it follows that $\Phi(\chi)=\Phi(\mathrm{Abstr}(\chi))$ and thus $\forall\alpha A(\Pi_0\chi,e(\Pi_0\chi))$.

The need for condition (2) is illustrated by the following example:

B.5.

Example III. Take $\mathcal{R}_x=\{U,B_x\}$, and let $\sqsubset$ hold in the cases $U\sqsubset U$, $B_x\sqsubset U$, $B_{x+1}\sqsubset B_x$. Then take $A(\alpha,x)$ to be

$$A(\alpha,x) \equiv \exists\chi\{\alpha=\Pi_0\chi \ \&\ [((\Pi_1\chi)0\equiv U \ \&\ x=0) \ \vee \ \vee\ ((\Pi_1\chi)0\equiv B_0 \ \&\ x=1)]\}.$$

Obviously $\forall\chi\exists!xA(\Pi_0\chi,x)$, for $\Pi_0\chi=\Pi_0\xi$ can be asserted only in case we know that χ,ξ are actually the same process; hence also $A(\Pi_0\xi,x)\ \&\ \Pi_0\xi=\Pi_0\chi \rightarrow A(\Pi_0\chi,x)$. However, it is also obvious that x cannot be found from an initial segment of $\Pi_0\chi$.

B.6.

Example IV. This example is included for its historical interest: it is the concept suggested by Brouwer's comments on choice sequences (there is more about this in sections A2, A6).

Countable decidable trees with all branches infinite are called spread laws in traditional intuitionistic terminology; the nodes of such a tree may be represented by finite sequences of natural numbers; if $\{n:bn\neq 0\}$ (b a lawlike sequence) is to represent the nodes of a spread law, b must satisfy Spr(b), where

$$\mathrm{Spr}(b) \equiv \forall nm(b(n*m)\neq 0 \rightarrow bn\neq 0) \; \& \; \forall n\exists x(bn\neq 0 \rightarrow b(n*\hat{x})\neq 0).$$

We define

$$\zeta\in b \equiv \forall x(b(\bar{\zeta}x)\neq 0); \; b\subset c \equiv \forall n(bn\neq 0 \rightarrow cn\neq 0).$$

As a first approximation, we may describe $\mathcal{R}_x$ as $\{R^a:\mathrm{Spr}(a)\}$, where $R^a\alpha \equiv \alpha\in a$, and $\sqsubset$ as corresponding to $\subset$ between spreads, i.e. if Spr(a), Spr(b), then $R^a \sqsubset R^b \Leftrightarrow a\subset b$. So a sequence is now pictured as

$$\langle x_0,a_0\rangle,\langle x_1,a_1\rangle,\ldots$$

where $\mathrm{Spr}(a_i), a_{i+1}\subset a_i$ for all i, and $\langle x_n\rangle_n\in a_i$ for all i; at each stage we restrict the sequences to a spread, which may become narrower as we proceed.

The picture is not quite correct according to our general schema. This can be repaired as follows.

(1) An arbitrary lawlike function a may be assumed to code a spread law $c=\Phi_0 a$ defined by

$$c0 = 1$$

$$c(n*\langle x\rangle) = \mathrm{sg}(cn)\cdot\mathrm{sg}(j_2a(n,x)+(1\dot{-}|x-j_1an|));$$

the part j_2a is used to enumerate permissible choices after n, the j_1a part ensures the continuation of n by at least one possible choice, i.c. j_1an.

(2) Subspreads of a given spread may be coded by an arbitrary lawlike sequence, as follows. If c' is a spread law, define d' by

$$d'j(n,x) = x\cdot c(n*\hat{x})+(1\dot{-}c(n*\hat{x}))\cdot c'n,$$

where $c'n = \min\{x:c(n*\hat{x})=1\}$.

If $c=\Phi_0 a$, d' is primitive recursive in a. For every n, $\lambda x.d'(n,x)$ enumerates all possible 'next choices'. Now define d primitive recursively from b,a by

$$d0=1,$$

$$d(n*\hat{x})=\mathrm{sg}(dn)\cdot\mathrm{sg}[c(n*\hat{x})\cdot\mathrm{sg}(j_2b(n,x)) + (1\dot{-}|x-d'(n,j_1bn)|)]$$

where $c=\Phi_0 a$. Let Φ denote the operator which constructs d from c,b, i.e. $d=\Phi(\Phi_0 a,b)$.

If we now think of the lawlike sequences (in the

intensional sense) as a decidable class,[†] we may put

$$\mathcal{R}_u \equiv_{\text{def}} \{R_{<a_0,\ldots,a_u>} : a_0,\ldots,a_u \text{ lawlike}\} \cup \{U\}$$

where

$$R_{<a_0,\ldots,a_u>} \equiv_{\text{def}} \lambda\chi.[\chi \in \Phi_u(a_0,\ldots,a_u)],\quad R_{<>} \equiv U,$$

$$\Phi_u(a_0,\ldots,a_u) \equiv \Phi(\Phi_{u-1}(a_0,\ldots,a_{u-1}),a_u),$$

and determine $\sqsubset$ by

$$R_{<a_0,\ldots,a_u>} \sqsubset R_{<b_0,\ldots,b_v>} \text{ iff } v+1=u \;\&\; \forall w \leq v(a_w \equiv b_w),$$

$$R_{<a_0,\ldots,a_u>} \sqsubset U,\quad U \sqsubset U.$$

By this choice of coding for spread laws and subspread laws relative to a given spread we have achieved decidability of $\mathcal{R}_n$ and $\sqsubset$.

3.7. *Projections of lawless sequences (in the sense of van Dalen and Troelstra (1970)).*

It is possible to represent various notions of sequences, such as are described in examples II,III for instance, by lawless sequences. To illustrate the idea, let us consider the approximation of the concept of example III by projections of lawless sequences. An arbitrary lawless sequence ε can be used to code a sequence $\Gamma\varepsilon$ of pairs: $<x_0,R_0>,<x_1,R_1>,<x_2,R_2>,\ldots$ (where the $R_i \in \mathcal{R}_i, R_{i+1} \sqsubset R_i$ as in example III) as follows:

$$(\Pi_0(\Gamma\varepsilon))(x) \equiv \begin{cases} (j_1\varepsilon)x \text{ if } (j_2\varepsilon)y=0 \text{ for all } y\leq x \\ sg(j_1\varepsilon)x \text{ if } (j_2\varepsilon)y\neq 0 \text{ for some } y\leq x \end{cases}$$

($sg0=0$, $sg(x+1)=1$).

$$(\Pi_1(\Gamma\varepsilon))(x) \equiv \begin{cases} U \text{ if } (j_2\varepsilon)y=0 \text{ for all } y\leq x \\ B_x \text{ if } (j_2\varepsilon)y\neq 0 \text{ for some } y\leq x. \end{cases}$$

Of course, such *projections* of lawless sequences should be regarded as distinct from the primitive concept represented, since a sequence according to example III is simply not *given* to us as obtained from a lawless sequence. But for

---

[†]I.e. a lawlike sequence is assumed to be *given* as such, hence we know whether the given object is a lawlike sequence or not.

mathematical purposes, especially in contexts where we do not refer in the language to the generation of the projected sequences from lawless sequences,[†] the projections can largely replace the primitive concepts. (See van Dalen and Troelstra (1970), Troelstra (1970)).

We can illustrate this by means of the following simple example. Let Ω be a continuous operator ('shift'-operator) on sequences defined by

$$(\Omega\zeta)(x) = \zeta(x+1).$$

Now consider

$$\mathfrak{A} = \{\Omega\varepsilon : \varepsilon \text{ lawless}\}.$$

Intuitively, we would expect $\mathfrak{A}$ to behave just like the universe of lawless sequences themselves; however $\forall\alpha\exists!x$-continuity does not hold:

$$\forall\alpha\exists!x(\exists\varepsilon(\alpha=\Omega\varepsilon \;\&\; \varepsilon 0 = x))$$

is obviously true (noting that from the axioms of lawless sequences one easily proves $\Omega\varepsilon=\Omega\eta \leftrightarrow \varepsilon=\eta$), although the x cannot be found from an initial segment of α.

However, if we restrict ourselves to the language of $\underset{\sim}{\mathrm{LS}}$, assuming the variables $\alpha,\beta,\gamma,\ldots$ to range over $\mathfrak{A}$, we can show that $\mathfrak{A}$ is in fact a model for $\underset{\sim}{\mathrm{LS}}$, proving by induction over the logical complexity of A ($A(\alpha,\beta_1,\ldots,\beta_n)$ a formula of $\underset{\sim}{\mathrm{LS}}$)

$$\underline{\forall}\varepsilon\in\hat{x}*n\; A(\Omega\varepsilon,\Omega\eta_1,\ldots,\Omega\eta_n) \leftrightarrow \underline{\forall}\varepsilon\in\hat{y}*n\; A(\Omega\varepsilon,\Omega\eta_1,\ldots,\Omega\eta_n)$$

(cf. Troelstra 1970, Thm.1)

and this can then be used (Troelstra 1970, Thm.4) to show that $\mathfrak{A}$ is a model for $\underset{\sim}{\mathrm{LS}}$.

To be able to treat Example IV by means of projections, we have to use lawless sequences of lawlike sequences, i.e. the elements are lawlike sequences. It can then be proved that the notions of Example II,III,IV all satisfy

$$\forall\alpha\exists x A(\alpha,x) \rightarrow \exists e\in K\forall\alpha A(\alpha,e(\alpha))$$

[†]E.g. if the projected sequences enter into the language as $\Pi_0(\Gamma\varepsilon)$ only. For instance, one may take the language of $\underset{\sim}{\mathrm{EL}}$ and assume $\forall a,\exists a$ to range over $\Pi_0(\Gamma\varepsilon)$, ε lawless.

and that Example IV moreover satisfies

$$A\alpha \rightarrow \exists a(\mathrm{Spr}(a) \ \& \ \alpha\in a \ \& \ \forall\beta\in aA\beta)$$

($A\alpha$ containing no choice parameters free except α).

B.8. A more rigorous treatment of example II (digression).

The study of the projections such as in van Dalen and Troelstra (1970) also suggests which axioms to assume for the notion of sequence of example II in order to obtain a rigorous refutation of $\forall\alpha\exists x$-continuity. The principal schemata we need are

A) Intensional continuity:

$$A(\chi,\chi_1,\ldots,\chi_n) \rightarrow \exists\sigma(\chi\in\sigma \ \& \ \forall\xi\in\sigma(\neq(\xi,\chi_1,\ldots,\chi_n) \rightarrow \rightarrow A(\xi,\chi_1,\ldots,\chi_n))),$$

where σ is an initial segment $\langle x_0,R_0\rangle,\ldots,\langle x_n,R_n\rangle$, and $\chi\in\sigma$ has the obvious meaning, i.e. $\exists x(\overline{\chi}x\equiv\sigma)$, and $\neq(\xi,\chi_1,\ldots,\chi_n)$ means that χ is a process distinct from $\xi_1,\ldots,\xi_n$.

B) Continuity: An operator Φ defined on all χ, with natural numbers as values, should assign a value to *all* sequences of pairs $\langle x_0,R_0\rangle,\langle x_1,R_1\rangle,\ldots$ (an assumption which has to be justified by the use of a process 'Abstr').

$R_i\in\mathcal{R}$, $R_{i+1}\sqsubset R_i$ etc. as in example II. Φ may be supposed to be coded by a neighbourhood function Φ^* acting on initial segments σ of the form $\langle x_0,R_0\rangle,\ldots,\langle x_m,R_m\rangle$:

$$\Phi^*\sigma\neq 0 \rightarrow \Phi^*\sigma=\Phi^*(\sigma*\tau) \quad (* \text{ denotes concatenation}),$$

$$\forall\chi\exists x(\Phi^*(\overline{\chi}x)\neq 0).$$

Then, in case (c) in section B.4, we may argue as follows: there exists, as shown by the argument under case (c), an operator Ψ such that

$$\forall\chi(\chi\in\sigma_0 \rightarrow \Pi_1\chi(\Psi\chi)\equiv N_{\Psi\chi}),$$

where $\sigma_0\equiv\langle\langle 0,B_0\rangle,\ldots,\langle 0,B_{y-1}\rangle\rangle$; let Ψ^* be the corresponding neighbourhood function, and let $\Psi^*\sigma=z+1$ for $\sigma \equiv \langle\langle 0,B_0\rangle,\ldots,\langle 0,B_x\rangle\rangle$. Without restriction of generality we may assume Ψ^* to be such that $\Psi^*\sigma=z+1 \rightarrow \mathrm{lth}(\sigma)>z$ for all σ, hence a contradiction follows, since $\Psi^*\sigma=z+1$ requires $B_z\equiv N_z$ for $z\leq x$.

B.9. The plausibility of continuity for the universe of all sequences.

The great variety of notions of choice sequence naturally leads to the question of whether there are some privileged notions. Certainly, the lawless sequences and the universes $\mathcal{U}, \mathcal{U}^*_\beta$ constructed from them deserve a special place. But it seems natural to ask whether we could not simply take the universe $\mathcal{U}^{**}$ of all sequences (i.e. embracing all intuitionistically meaningful concepts of sequence, from lawlike to lawless) and show that it satisfies some continuity schema. $\mathcal{U}^{**}$ is certainly closed under continuous operations, and satisfies the bar theorem BI_D (since it contains the lawless sequences). However, justifying e.g. $\forall\alpha\exists x$-continuity presents more difficulties. A plausibility argument might run as follows.

$\forall\alpha\exists x A(\alpha,x)$ requires the existence of an operator Φ such that $\forall\alpha A(\alpha,\Phi\alpha)$; Φ conceivably uses intensional information about the arguments for its computation. However, since Φ should be defined for *all* elements of $\mathcal{U}^{**}$, including the lawless sequences, we cannot assume, for an arbitrary sequence, at any given stage, more than an initial segment to be known. Hence Φ should be continuous. This reasoning is not entirely cogent however, as is shown by example I in section B.2: if the universe is just the union of lawlike and lawless sequences, we can always tell which of the two is the case, and this decision is not made by looking at initial segments (i.e. some intensional information apart from initial segments is always available: the sequence is lawlike or lawless).

On the other hand, $\mathcal{U}^{**}$ contains also all sequences which are intermediate between lawless or lawlike (i.e. we do not know whether they will ultimately conform to a law or not). This class is so 'diffuse' (with no sharp boundaries between the possible concepts) that we may argue that for a functional Φ defined on *all* elements of $\mathcal{U}^{**}$ we cannot expect that *definite* intensional information about the argument, on which the value of Φ depends in an essential way, will always be available. This argument resembles an appeal to impotency: it seems impossible to extract *always* definite information except from initial segments; hence, since Φ must be defined everywhere on $\mathcal{U}^{**}$ we can only think of Φ as acting on initial segments (of numerical values). This is analogous to the argument for the uniformity principle UP for sets in Troelstra (1973A), which states (X ranging over sets of natural numbers)

UP $\quad \forall X \exists x A(X,x) \to \exists x \forall X A(X,x)$.

So, for $\mathcal{U}^{**}$ we can (at least at present) at most justify $\forall\alpha\exists x$-continuity by a plausibility argument, whereas for e.g. lawless sequences we have actually *derived* (informally, but rigorously) continuity.

B.10. Continuity and Church's thesis.

Church's thesis forcefully emphasizes intensional aspects of sequences. We have already seen how CT refutes $\forall\alpha\exists x$-continuity (section 5.6). On the other hand, a system such as $\underset{\sim}{\underset{\sim}{\mathrm{EL}}}_1$ + CT is consistent with $\forall\alpha\exists!x$-continuity, thereby providing a simple proof of the formal independence of $\forall\alpha\exists x$-continuity from $\forall\alpha\exists!x$-continuity (see Troelstra (1973), 2.6.17, 3.6.14(i), 3.2.18, 3.2.22). However, the proof of $\forall\alpha\exists!x$-continuity in a suitable extension of $\underset{\sim}{\underset{\sim}{\mathrm{EL}}}_1$ + CT shows that continuity here holds for quite different reasons as in the case of choice sequences: the principal reason being that the condition of assigning the same value to all extensionally equal recursive functions is so strong.

APPENDIX C

A NOTION OF CHOICE SEQUENCE SATISFYING CS

C.1.

In this appendix we describe a notion of choice sequence which was originally devised to satisfy CS (see the remarks in sections A.7 and 5.1). The account below is a more careful and critical version of the discussion in Troelstra (1968, 1969, 1969B). Our reasons for including a discussion of this notion at all are

(i) to see how far we can get in 'approximating' CS (i.e. we claim a certain pedagogical interest) and, more important

(ii) because it suggests a still more flexible method of constructing universes of sequences from lawless sequences, which may be a starting point for further research.

C.2. Description of the notion: GC-sequences.

('GC' stands for 'Generated by Continuous operations'.) We think of a choice sequence α as started by generating values $\alpha 0, \alpha 1, \ldots$ - then, at some stage we decide to make α dependent on another, 'fresh' sequence α_0 by means of a continuous operation, i.e. $\alpha_0 = \Gamma_0 \alpha_0$ ($\Gamma_0 : N^N \to N^N$) ; from then on, α is determined by choosing values of α_0 - at a later stage, we may in turn wish to make α_0 dependent on another sequence α_1, so $\alpha_0 = \Gamma_1 \alpha_1$, etc.

To express this in a way which is in agreement with the descriptional schema introduced in section B.3 for generalizations of lawless sequences, we can also describe GC-sequences as a process of choosing pairs

$$\langle x_0, R_0 \rangle \ \langle x_1, R_1 \rangle \ \langle x_2, R_2 \rangle, \ldots ,$$

where the R_i are now conditions of the form

(1) $\quad \alpha = \Gamma_0 \alpha_0 \ \& \ \alpha_0 = \Gamma_1 \alpha_1 \ \& \ldots \& \ \alpha_{n_i - 1} = \Gamma_{n_i} \alpha_{n_i}$

or of the form

(1') $\quad \alpha = \alpha$ (the universal condition U, i.e. no restriction).

We also have to specify the relation $\sqsubset$:

$U \sqsubset U$, $U \sqsubset R_i$ if R_i is of the form (1),
$R_i \sqsubset R_i$, and $R_i \sqsubset R_j$ if R_j can be written as
R_i & $\alpha_{n_i} = \Gamma_{n_i+1}\alpha_{n_i+1}$ &...& α_{n_j-1} $\Gamma_{n_j}\alpha_{n_j}$ $(j>i)$.

Note that the R_i are now not any longer 'lawlike'; they not only involve α (the sequence to be generated) but also other choice sequences as parameters.

At first sight, it seems as if a condition of the form (1) could be summed up by saying that α is in the range of the continuous operator $\Gamma_0\Gamma_1\ldots\Gamma_{n_i}$. However, this is not sufficient; the 'individuality' of α_{n_i} does matter, as we shall see in the discussion below. To see that reducing the matter to conditions of the form 'α is in a certain range' is actually incorrect† (for our purposes), note that 'α is in the range of Γ' is a lawlike condition (say R_Γ). To specify $\sqsubset$ for such conditions, the natural thing would be to take $R_\Gamma \sqsubset R_{\Gamma'}$ iff $\exists\Gamma''(\Gamma=\Gamma'\Gamma'')$. However, the same criticism would apply to such a notion of choice sequence as to the examples described in Appendix B: they would in general not be closed under non-trivial continuous operations!

To see that our universe, as just described, *is* closed under continuous operations, consider any choice sequence α, which starts with

$$\langle x_0,R_0\rangle,\langle x_1,R_1\rangle,\ldots .$$

Then $\beta=\Gamma\alpha$ can be described as a choice sequence starting with

$$\langle y_0,R_0'\rangle,\langle y_1,R_1'\rangle,\langle y_2,R_2'\rangle,\ldots ,$$

where $y_i=(\Gamma\alpha)(i)$, and R_i' can be written as

$$R_i'(\beta) \equiv [\beta=\Gamma\alpha \ \& \ R_i(\alpha)].$$

Here it is the proper place to make another comment. So far we have presented a simplified picture, in as much as we omitted to take into account the possibility that a choice

---

†For this reason the proposal in Kreisel (1968), Postscript (page 247), for a notion satisfying CS is inadequate.

sequence is obtained from two or more other choice sequences; i.e. the R_i should also contain clauses of the form

$$\alpha_k = \Gamma_k \nu_{r(k)}(\alpha_{k+1,1},\ldots,\alpha_{k+1,r(k)}).$$

This makes the general form of the conditions R_i much more complicated; they now look something like:

$$\alpha = \Gamma_0 \nu_{r(0)}(\alpha_{0,0},\ldots,\alpha_{0,r(0)}) \;\&$$

$$\bigwedge_{0 \le i \le r(0)} \alpha_{0,i} = \Gamma_{1,i} \nu_{r(1,i)}(\alpha_{1,i,0},\ldots,\alpha_{1,i,r(1,i)}) \;\&$$

$$\bigwedge_{0 \le i \le r(0)} \; \bigwedge_{0 \le j \le r(1,i)} \alpha_{1,i,j} =$$

$$= \Gamma_{2,i,j} \nu_{r(2,i,j)}(\alpha_{2,i,j,0},\ldots,\alpha_{2,i,j,r(2,i,j)}) \;\&$$

...

... .

Some simplifications are possible, but the notation looks pretty horrible. Therefore, for simpler notation, we shall restrict ourselves to the simpler case. What we are generating in this case, in contrast to the simpler cases discussed in Appendix B is really a 'network' of choice sequences instead of a single one. It should be stressed also that for a choice sequence α starting with $\langle x_0,R_0\rangle,\ldots,\langle x_i,R_i\rangle$, with R_i as indicated in formula (1), all freedom of continuation for α is given by the freedom of continuation for the α_{n_i}. With all these provisos in mind, we shall now try to establish some principles valid for the notion as just described arguing more or less in analogy with the reasoning in Chapter 2 and Appendix B.

C.3. Continuity schemas for GC-sequences.

We may argue largely along the lines used in Appendix B, with a slight difference.
Assume for our sequences

$$\forall\alpha\exists x A(\alpha,x).$$

(A being extensional with respect to α). Using a selection principle, there is an operator Ψ (possibly using intensional information about our sequences) such that $A(\alpha,\Psi\alpha)$. Now con-

sider any α starting with $\langle x_0,R_0\rangle,\ldots,\langle x_i,R_i\rangle,\ldots$ and apply a process 'Abstr' (similar to the ones used before in Chapter 2 and Appendix B) to obtain a sequence $\langle x_0,U\rangle,\ldots,\langle x_i,U\rangle,\ldots$ 'indistinguishable' (for Ψ) from a GC-sequence.
Assume Ψ to be given on such sequences Abstr(α) by a continuous (extensional) operator Γ, represented by a neighbourhood function e, i.e.

$$\forall\alpha A(\text{Abstr}(\alpha),e(\alpha)).$$

Suppose Ψ for Abstr(α) to be determined from the initial segment $\langle x_0,U\rangle,\ldots,\langle x_n,U\rangle$, and assume Ψ to be determined for α from

$$\langle x_0,R_0\rangle,\ldots,\langle x_m,R_m\rangle;$$

without restriction we may assume $m\geq n$.
In the case when $R_m\equiv U$, then $\Psi(\alpha)=\Psi(\text{Abstr}(\alpha))$, without a problem. So now assume $R_m\not\equiv U$, i.e. R_m is of the form

$$R\alpha \equiv [\alpha=\Gamma_0\alpha_0 \ \&\ldots\&\ \alpha_{i-1}=\Gamma_i\alpha_i].$$

Then for a sequence β starting with

$$\langle x_0,U\rangle,\ldots,\langle x_n,U\rangle,\langle x_{n+1},R\rangle \ (x_{n+1}=\alpha(n+1))$$

also $\Psi\beta=\Psi(\text{Abstr}(\alpha))$. However, $\beta=\alpha$, since both satisfy R, and $R\alpha$ & $R\alpha' \to \alpha=\alpha'$ (α,α' being (extensionally) completely determined by the parameter α_i occurring in R), and therefore

$$\forall\alpha A(\alpha,\Psi(\text{Abstr}(\alpha)))$$

i.e.

$$\forall\alpha A(\alpha,e(\alpha)).$$

Thus we have justified

$$\forall\alpha\exists x A(\alpha,x) \to \exists e\in K_{CS}\forall\alpha A(\alpha,e(\alpha)),$$

K_{CS} being the class of neighbourhood functions with respect to the universe of GC-sequences as described above.

The same type of argument does *not* justify continuity in the form

CONT$_1$ $\quad \forall\alpha\exists\beta A(\alpha,\beta) \to \exists\Gamma\forall\alpha A(\alpha,\Gamma\alpha)$

(Γ a variable for lawlike continuous operations of type $N^N \to N^N$). To see this, one should realize that the very idea of Abstr(α) means that any initial segment of this sequence is indistinguishable from a choice sequence:

Abstr(α) *is* a choice sequence if we abstract from the fact that we originally decided to construct it by always taking U for the restricting sequence. As soon as this fact is used somewhere, we are really thinking of Abstr(α) in a different way (as a lawless sequence, in fact) because we are admitting types of information incompatible with the assumption that Abstr(α) is a choice sequence of the kind described.

Therefore the assumption that CONT_1 should hold can only be interpreted (as long as no better analysis is available) as imposing a certain restriction on the interpretation of the quantifier combination $\forall\alpha\exists\beta$ which is not (in any obvious way) already implicit in the intended meaning of this quantifier combination for choice sequences. On the other hand, as we have seen from the discussions above, $\forall\alpha\exists\beta$-continuity fails to be justified by the arguments leading to $\forall\alpha\exists x$-continuity, i.e. it appears as if $\forall\alpha\exists\beta$-continuity represents a slight strengthening of the intended interpretation. In this connection it should be noted that the 'obvious' type of counterexample to $\forall\alpha\exists\beta$-continuity as given by Myhill (see section A.6) fails precisely because we cannot show $\forall\alpha\exists\beta A(\alpha,\beta)$ by constructing non-continuously from α, unless we can show the universe of sequences to be *closed* under such non-continuous operations. The situation with respect to CONT_1 is not unlike 'Church's thesis' for lawlike sequences, or 'Brouwer's dogma' for neighbourhood functions.

C.4. The principle of analyticity.

Suppose we have a proof of $A\alpha$ (A not containing other choice parameters besides α, and extensional with respect to α). Then the proof can only make use of the information concerning α available at a certain stage, i.e. an initial segment

$$\langle x_0, R_0\rangle, \ldots, \langle x_n, R_n\rangle,$$

and suppose $R_n(\alpha)$ to be of the form

$$\alpha=\Gamma_0\alpha_0 \ \&\ldots\& \ \alpha_{i-1}=\Gamma_i\alpha_i.$$

Suppose first, for simplicity, that at the stage n α_i has just been introduced and is still completely 'fresh', i.e. no values have been determined. Then for any continuation (given by providing values and conditions on α_i) A will hold; α_i is as yet completely undetermined. Therefore, if we put $\Gamma_0\Gamma_1\ldots\Gamma_i\equiv\Gamma$, then $\alpha=\Gamma\alpha_i$, and since α_i is 'fresh' $\forall\beta A(\Gamma\beta)$. Expressed in a schema:

$$A\alpha \to \exists\Gamma[\exists\beta(\alpha=\Gamma\beta) \ \& \ \forall\gamma A(\Gamma\gamma)].$$

(We have chosen to disregard the fact that for α_i we cannot substitute $\alpha_0,\alpha_1,\ldots,\alpha_{i-1}$ or α itself, i.e. we have assumed that quantifying over all γ amounts to the same (with respect to A) as quantifying over all γ except finitely many individuals).

If at stage n we have determined an initial segment m of α_i, we refine the argument as follows. Let $\Gamma^{(m)}$ be the continuous operator of type $N^N \to N^N$ such that

$$(\Gamma^{(m)}\alpha)(x)=(m)_x \ \text{ for } x<\mathrm{lth}(m)$$
$$(\Gamma^{(m)}\alpha)(x)=\alpha x \ \text{ for } x\geq\mathrm{lth}(m).$$

Now A should hold for any consistent extension of $\langle x_0,R_0\rangle,\ldots,\langle x_n,R_n\rangle$, hence also for a β starting with

$$\langle x_0,R_0\rangle,\ldots,\langle x_n,R_n\rangle,\langle x_{n+1},R_{n+1}\rangle,$$

where $x_{n+1}=\alpha(n+1)$ and where R_{n+1} is of the form, for suitable i

$$R_n\alpha \ \& \ \alpha_i=\Gamma^{(m*p)}\alpha_{i+1},$$

α_{i+1} being completely undetermined at stage $n+1$, and $m*p$ being the minimal extension of m which guarantees that $x_{n+1}=(\Gamma_0\Gamma_1\ldots\Gamma_i\Gamma^{(m*p)}\beta)(n+1)$ for all β. (Here 'minimal' refers to the neighbourhood functions by which $\Gamma_0,\ldots,\Gamma_i$ are given, i.e. $m*p$ must be the least initial segment ($\succeq m$) from which these neighbourhood functions define a value x_{n+1}). Then as before

$$\exists\beta(\alpha=\Gamma\beta) \ \& \ \forall\gamma A(\Gamma\gamma)$$

for $\Gamma\equiv\Gamma_0\Gamma_1\ldots\Gamma_i\Gamma^{(m*p)}$.

C.5. *Discussion of the analysis.*

The principal defect of the analysis is in the complexity of the notion itself - it appears too much as an ad hoc device to find a notion validating CS. More specific weaknesses are

(i) The informal analysis might gain in rigour if a more adequate notational apparatus would be available for discussing 'networks' of sequences; at times it is misleading to talk about an individual sequence $\langle x_0, R_0\rangle, \langle x_1, R_1\rangle, \ldots$ whereas we are really dealing with 'networks'.

(ii) There is a (minor) additional assumption in the principle of analyticity, as may be seen from the discussion in section C.4.

(iii) The justification of $\forall\alpha\exists\beta$-continuity is not complete; at present we can only interpret this as a (slight) strengthening of the intended interpretation of the quantifier combination $\forall\alpha\exists\beta$.

With respect to the 'Brouwerian dogma' (amounting to $K_{\text{CS}}=K$ in this case) the situation here does not essentially differ from the situation for lawless sequences.

C.6. *Outline of a research project.*

The preceding discussion of GC-sequences suggests the possible interest of an extension of the treatment of projections of lawless sequences (in the sense of van Dalen and Troelstra 1970) so as to be able to approximate choice sequences which can have non-trivial relationships to each other (e.g. if they are closed under certain non-trivial continuous operations).

Let us consider, as one of the simplest examples illustrating the possibilities, the notion of free sequence (first mentioned in Troelstra 1968). The information available at any stage for a free sequence is (1) an initial segment of values, (2) that the sequence is extensionally equal to certain other free sequences (*not given* from the start as being identical with the sequence considered).

At each stage, we choose a value and (possibly) require extensional equality with certain other free sequences (when we can consistently do so because of agreement of initial segments already chosen).

In carrying out the program of approximating this notion by means of projections, we encounter a prima facie difficulty: the natural procedure for generating a free sequence would be to let the choice of values as well as identifications with other free sequences be governed by (ordinary) lawless sequences. However, how can we indicate which free sequences to identify if we have no 'names' for them? This problem can be solved by the observation in section 3.19, that we can in fact construct a countable model for the theory $\underset{\sim}{LS}$ from a single lawless sequence.

So an attempt to build a model for the theory of free sequences might look as follows: let α,β be a pair of independent lawless sequences. We construct a countable universe $\varepsilon_0,\varepsilon_1,\varepsilon_2,\ldots$ approximating the free sequences by letting $\alpha^n = n*(\alpha)_n$ (cf. section 3.19 for notation) govern the values of ε_n, and let β govern the identifications between the ε_n.

For example, consider stage m: if $\beta m=0$, no new identifications are required. If $\beta m>0$, we consider m_0,m_1 such that $\beta m=(2m_0+1)2^{m_1}$; and if $\bar{\varepsilon}_{m_0}(m)=\bar{\varepsilon}_{m_1}(m)$, they are henceforth to be identified. Let, for any n, $\{\varepsilon_{n_1},\varepsilon_{n_2},\ldots,\varepsilon_{n_k}\}$ be the set with which ε_n has to be identified according to $\bar{\beta}(m+1)$; assume $n_1<n_2<\ldots<n_k$. Then $(\varepsilon_n)(m)=(\alpha^{n_1})(m)$.

For this proposed model one can then try to establish the validity of some schemata which one intuitively expects to hold for free sequences, such as

$$(1)\qquad \begin{cases} A\alpha \rightarrow \exists n[\alpha\in n \;\&\; \forall\beta\in n A\beta] \\ \forall\alpha\exists x A(\alpha,x) \rightarrow \exists e\forall\alpha A(\alpha,e(\alpha)) \end{cases}$$

(α the only 'choice' variable free in A); A must belong to a restricted language (e.g. of $\underset{\sim}{LS}$).

It would undoubtedly be very instructive if we could treat e.g. free sequences by projections and establish the schemata (1) as being valid for such projections. If we were successful in this, then presumably a much more complex schema of the type outlined above, involving lawless sequences of elements of K might be constructed to approximate GC-sequences (although at present we have little reason to prefer an approximation of GC-sequences over many other possibilities). A quite different possibility would be the study of lawless sequences and choice sequences of higher types, as suggested by bar recursion of higher type (cf. Troelstra 1973 1.9.26).

BIBLIOGRAPHY

An extensive bibliography on the metamathematics of intuitionistic analysis and choice sequences up till 1965 is found in Kleene and Vesley (1965).

All papers of Brouwer listed below are reprinted in the following edition of the collected works:

Brouwer, L.E.J. (1975), *Collected Works I, Philosophy and Foundations of Mathematics* (ed. A.Heyting). North-Holland, Amsterdam and American Elsevier, New York.

We shall refer to this volume, in which papers in Dutch have been translated into English, as 'Brouwer, Collected works'.

Abbreviations:

IPT : *Intuitionism and proof theory* (ed. A. Kino, J. Myhill and R.E. Vesley) North-Holland, Amsterdam.

JSL : The Journal of Symbolic Logic.

KNAW: Koninklijke Nederlandse Akademie van Wetenschappen.

LMPS: Logic, Methodology and Philosophy of Science.

KNAW Proc: Koninklijke Nederlandse Akademie van Wetenschappen Proceedings of the section of sciences.

Aczel, P.H.G. (1968). Saturated intuitionistic theories. In *Contributions to mathematical logic* (ed. H. Arnold Schmidt, K. Schütte and H.-J. Thiele), pp. 1-11. North-Holland, Amsterdam.

Beth, E.W. (1956). Semantic construction of intuitionistic logic. *Mededelingen der KNAW* N.S. 19, no.11.

——— (1959). *The foundations of mathematics*. North-Holland, Amsterdam. Second, revised edition 1965, reprinted 1968.

Bishop, E. (1967). *Foundations of constructive analysis*. McGraw-Hill, New York.

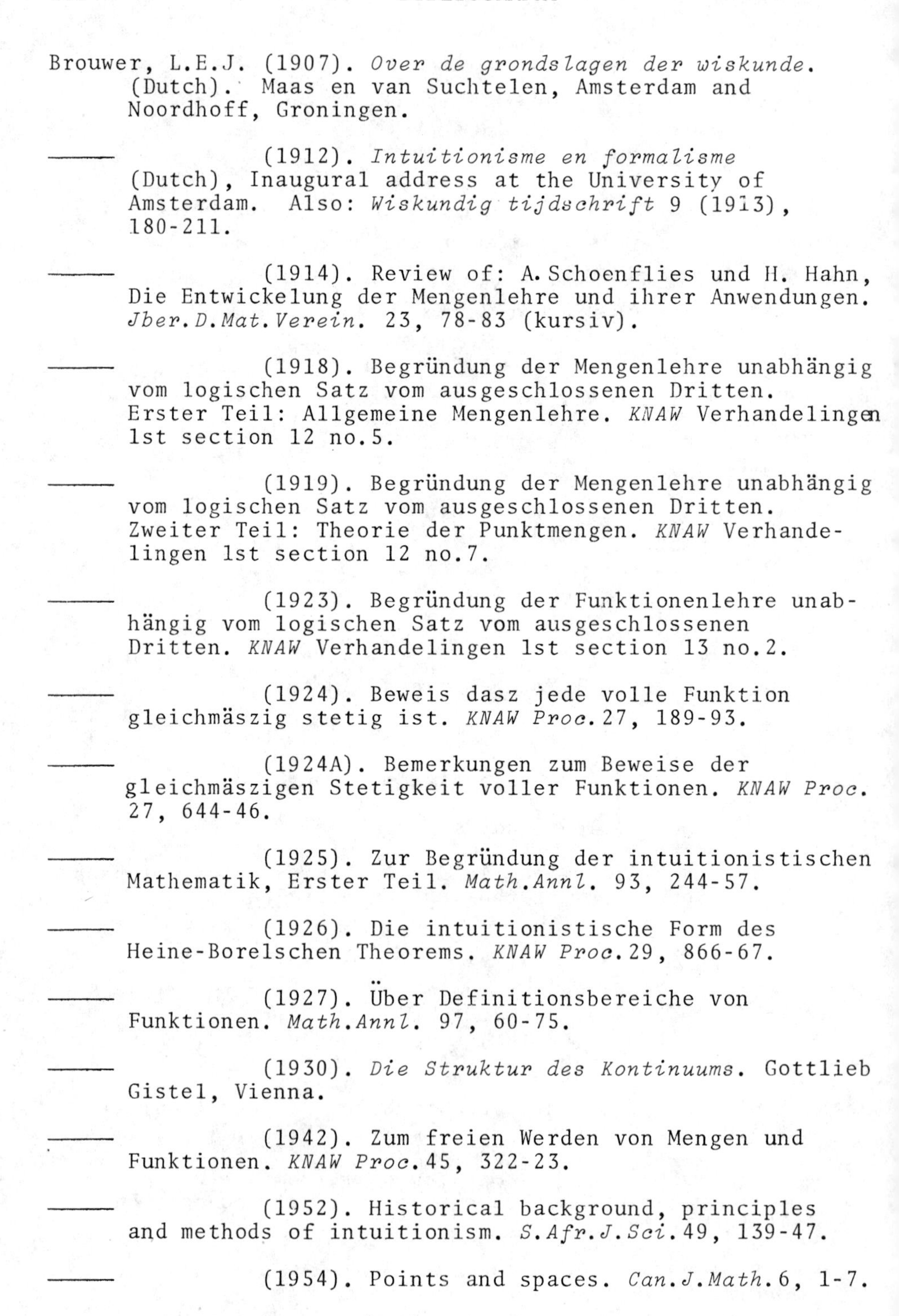

Brouwer, L.E.J. (1907). *Over de grondslagen der wiskunde.* (Dutch). Maas en van Suchtelen, Amsterdam and Noordhoff, Groningen.

——— (1912). *Intuitionisme en formalisme* (Dutch), Inaugural address at the University of Amsterdam. Also: *Wiskundig tijdschrift* 9 (1913), 180-211.

——— (1914). Review of: A. Schoenflies und H. Hahn, Die Entwickelung der Mengenlehre und ihrer Anwendungen. *Jber.D.Mat.Verein.* 23, 78-83 (kursiv).

——— (1918). Begründung der Mengenlehre unabhängig vom logischen Satz vom ausgeschlossenen Dritten. Erster Teil: Allgemeine Mengenlehre. *KNAW* Verhandelingen 1st section 12 no.5.

——— (1919). Begründung der Mengenlehre unabhängig vom logischen Satz vom ausgeschlossenen Dritten. Zweiter Teil: Theorie der Punktmengen. *KNAW* Verhandelingen 1st section 12 no.7.

——— (1923). Begründung der Funktionenlehre unabhängig vom logischen Satz vom ausgeschlossenen Dritten. *KNAW* Verhandelingen 1st section 13 no.2.

——— (1924). Beweis dasz jede volle Funktion gleichmäszig stetig ist. *KNAW Proc.* 27, 189-93.

——— (1924A). Bemerkungen zum Beweise der gleichmäszigen Stetigkeit voller Funktionen. *KNAW Proc.* 27, 644-46.

——— (1925). Zur Begründung der intuitionistischen Mathematik, Erster Teil. *Math.Annl.* 93, 244-57.

——— (1926). Die intuitionistische Form des Heine-Borelschen Theorems. *KNAW Proc.* 29, 866-67.

——— (1927). Über Definitionsbereiche von Funktionen. *Math.Annl.* 97, 60-75.

——— (1930). *Die Struktur des Kontinuums.* Gottlieb Gistel, Vienna.

——— (1942). Zum freien Werden von Mengen und Funktionen. *KNAW Proc.* 45, 322-23.

——— (1952). Historical background, principles and methods of intuitionism. *S.Afr.J.Sci.* 49, 139-47.

——— (1954). Points and spaces. *Can.J.Math.* 6, 1-7.

van Dalen, D. (1973). Lectures on intuitionism. In *Cambridge summer school in mathematical logic* (ed. H. Rogers and A.R.D. Mathias), pp.1-89. (Lecture notes in mathematics 337) Springer-Verlag, Berlin.

——— (1974). Choice sequences in Beth-models. Locos 20, Department of Mathematics, University of Utrecht. Contained in revised form in van Dalen (1975A).

——— (1975). Experiments in lawlessness. Locos 28, Department of Mathematics, University of Utrecht. Contained in revised form in van Dalen (1975A).

——— (1975A). An interpretation of intuitionistic analysis. Preprint 14. Department of Mathematics, University of Utrecht. To appear in *Ann.Math.Logic*.

van Dalen, D. and Troelstra, A.S. (1970). Projections of lawless sequences. In *IPT*, pp. 163-86.

Davis, M. (1965). *The undecidable*. Raven Press, Hewlett.N.Y.

Dragalin, A.G. (1973). Constructive mathematics and models of intuitionistic theories. In *LMPS* IV, (ed. P. Suppes, L. Henkin, Gr.C. Moisil and A. Joja), pp. 111-28. North-Holland, Amsterdam.

Dummett, M.A.E. (1977). *Elements of intuitionism*. Clarendon Press, Oxford.

Dyson, V.H. and Kreisel G. (1961). Analysis of Beth's semantic construction of intuitionistic logic. Technical Report 3, Applied Mathematics and Statistics Laboratories, Stanford University.

Fitting, M.C. (1969). *Intuitionistic logic, model theory and forcing*. North-Holland, Amsterdam.

Gielen, W. (1969). In *Abstracts of the 5th Dutch mathematical congress (5e Nederlands mathematisch congres)*, Wageningen 10-11 April 1969.

Goodman, N.D. (1968). Intuitionistic arithmetic as a theory of constructions (thesis, Stanford University).

Heyting, A. (1930). Die formalen Regeln der intuitionistischen Mathematik III. *Sber.Preuss.Akad.Wiss.*,158-69.

——— (1956). *Intuitionism, an introduction* (second revised edition 1966, third revised edition 1972). North-Holland, Amsterdam.

Howard, W.A. and Kreisel, G. (1966). Transfinite induction and bar induction of types zero and one, and the role of continuity in intuitionistic analysis. *JSL* 31, 325-58.

Kleene, S.C. (1952). Recursive functions and intuitionistic mathematics. *Int.Congr.Math.* 11, I 679-85.

——— (1957). Realizability. In *Summaries of talks presented at the Summer Institute of Symbolic Logic in 1957, at Cornell University*, vol.1, pp.100-104 (2nd edition 1960. Princeton N.J.). Communications Research Division, Institute for Defense Analyses. Reprinted (1959) in: *Constructivity in mathematics* (ed. A. Heyting), pp. 258-89. North-Holland, Amsterdam.

——— (1965). Classical extensions of intuitionistic mathematics. In *LMPS* (ed. Y. Bar-Hillel) (being the proceedings of the 1964 International Congress held at Jerusalem Aug.26 - Sept.2), pp. 31-44. North-Holland, Amsterdam.

——— (1965A). Logical calculus and realizability. *Acta Philos.Fenn.*18, 137-44.

——— (1968). Constructive functions in "The Foundations of Intuitionistic Mathematics". In *LMPS* III (ed. B. van Rootselaar and J.F. Staal), pp. 137-44. North-Holland, Amsterdam.

——— (1969). Formalized recursive functionals and formalized realizability. *Mem.Amer.Math.Soc.*89, Providence (Rh.I.).

Kleene, S.C. and Vesley, R.E. (1965). *The foundations of intuitionistic mathematics, especially in relation to recursive functions*. North-Holland, Amsterdam.

Kreisel, G. (1958). A remark on free choice sequences and the topological completeness proofs. *JSL* 23, 369-88.

——— (1962). Weak completeness of intuitionistic predicate logic. *JSL* 27, 139-58.

——— (1963). Section IV in: Stanford report on the foundations of analysis (Stanford University, mimeographed).

——— (1965). Mathematical logic. In *Lectures on modern mathematics vol.III* (ed. T.L. Saaty), pp. 95-195. John Wiley and Sons, New York.

——— (1966). Review of Kleene and Vesley (1965), *JSL* 31, 258-61.

——— (1967). Informal rigour and completeness proofs. In *Problems in the philosophy of mathematics* (ed. I. Lakatos), pp. 138-86. North-Holland, Amsterdam.

——— (1968). Lawless sequences of natural numbers. *Compositio Math.*20, 222-48.

Kreisel, G. (1970). Church's thesis: A kind of reducibility axiom for constructive mathematics. In *IPT*, pp. 121-50.

Kreisel, G. and Troelstra, A.S. (1970). Formal systems for some branches of intuitionistic analysis. *Ann.Math. Logic* 1, 229-387.

Kripke, S.A. (1965). Semantical analysis of intuitionistic logic I. In *Formal systems and recursive functions* (ed. J.N. Crossley and M.A.E. Dummett), pp. 93-130. North-Holland, Amsterdam.

Lopez-Escobar, E.G.K. and Veldman, W. (1975). Intuitionistic completeness of a restricted second-order logic. In *Proof theory symposion Kiel 1974* (ed. J. Diller and G.H. Müller) pp. 198-232. (Lecture notes in mathematics 500), Springer-Verlag, Berlin.

Luckhardt, H. (1970). Ein Henkin-Vollständigkeitsbeweis für die intuitionistische Prädikatenlogik bezüglich der Kripke-Semantik. *Arch.Math.Logik Grundforsch.* 13, 55-9.

Minc, G.E. (1975). Finitnoe issledovanie transfinitnyh vyvodov ('Finite investigation of infinite derivations') (Russian, with English summary). *Zap. Naučn. Sem. Leningrad. Otdel. Mat.Inst.Steklov. (LOMI)* 49, 67-122. To be translated in *J.Soviet Math.*

Moschovakis, J.R. (1971). Can there be no non-recursive functions? *JSL* 36, 309-15.

——— (1973). A topological interpretation of second-order intuitionistic arithmetic.

Myhill, J. (1967). Notes towards an axiomatization of intuitionistic analysis. *Logique et Analyse* (N.S.) 9, 280-97.

——— (1968). Formal systems of intuitionistic analysis I. In *LMPS III* (ed. B. van Rootselaar, J.F. Staal), pp. 161-78. North-Holland, Amsterdam.

——— (1970). Formal systems of intuitionistic analysis II. In *IPT*, pp. 151-62.

Putman, H. (1957). Arithmetic models for consistent formulae of quantification theory (abstract). *JSL* 22, 110-11.

van Rootselaar, B. (1952). Un problème de M. Dijkman. *Indag.math.* 14, 405-7.

Schütte, K. (1968). *Vollständige Systeme modaler und intuitionistischer Logik*, Springer-Verlag, Berlin.

Scott, D.S. (1968). Extending the topological interpretation to intuitionistic analysis. *Compositio math.*20, 222-48.

——— (1970). Extending the topological interpretation to intuitionistic analysis II. In *IPT*, pp. 235-55.

de Swart, H. (1974). Another intuitionistic completeness proof. Report, Mathematisch Instituut, Katholieke Universiteit Nijmegen. (Submitted for publication to *JSL*.)

Thomason, R.H. (1968). On the strong semantical completeness of the intuitionistic predicate calculus, *JSL* 33, 1-7.

Troelstra, A.S. (1967). Intuitionistic continuity. *Nieuw Archc.Wisk.* Third series 15, 2-6.

——— (1968). The theory of choice sequences. In *LMPS III* (ed. B. van Rootselaar and J.F. Staal), pp. 201-23. North-Holland, Amsterdam. (Some corrections in Troelstra 1973.)

——— (1968A). The use of Brouwer's principle in intuitionistic topology. In *Contributions to mathematical logic* (Proceedings of the logic colloquium Hannover 1966) (ed. H. Arnold Schmidt, K. Schütte and H. -J. Thiele), pp. 289-98. North-Holland, Amsterdam.

——— (1969). *Principles of intuitionism.* (Lecture notes in mathematics 95) Springer-Verlag, Berlin. (Some corrections in Troelstra 1973.)

——— (1969A). Notes in the intuitionistic theory of sequences I. *Indag.math.*31, 430-40.

——— (1969B). Informal theory of choice sequences. *Studia logica* 25, 31-52.

——— (1970). Notes on theintuitionistic theory of sequences II. *Indag.math.*32, 99-109.

——— (1970A). Notes on the intuitionistic theory of sequences III. *Indag.math.*32, 245-52.

——— (1971). An addendum. *Ann.Math.Logic* 3, 437-39.

——— (1973). Chapters I-IV in *Metamathematical investigation of intuitionistic arithmetic and analysis* (ed. A.S. Troelstra) (Lecture notes in mathematics 344) Springer-Verlag, Berlin.

Troelstra, A.S. (1973A). Notes on intuitionistic second order arithmetic. In *Cambridge summer school in mathematical logic* (ed. A.R.D. Mathias and H. Rogers), pp. 171-205. (Lecture Notes in Mathematics 337) Springer-Verlag, Berlin.

——— (1974). Note on the fan theorem. *JSL* 39, 594-96.

——— (1974A). Markov's principle and Markov's rule for theories of choice sequences. Report 74-12, Department of Mathematics, University of Amsterdam. Final version appeared in *Proof theory symposion Kiel 1974* (ed. J. Diller and G.H. Müller) (Lecture notes in mathematics 500), pp. 370-83. Springer-Verlag, Berlin (1975).

——— (1975). Non-extensional equality. *Fundam. Math.* 82, 307-22.

——— (1976). Completeness and validity of intuitionistic predicate logic. Report 76-05, Department of Mathematics, University of Amsterdam. To appear in the Proceedings of the European meeting of the Association for Symbolic Logic at Clermont-Ferrand, July 1975.

——— (1976A). Two notes on intuitionistic analysis. Report 76-04, Department of Mathematics, University of Amsterdam. (The first note is an appendix to van Dalen (1975A), the second note is to appear in *Indag.math.*)

Vaught, R.E. (1960). Sentences true in all constructive models. *JSL* 25, 39-53.

Vesley, R.E. (1970). A palatable alternative to Kripke's schema. In *IPT*, pp. 197-207.

INDEX

List of formal systems

List of axioms and axiom-schemata.

Other notations of more than local significance are either standard or to be found on pages 7-10, 15, 31-33, 101, 106, 107.